AF556198

Traditional Forestry Management

Traditional Forestry Management

Ajay Mukherjee

RANDOM PUBLICATIONS
NEW DELHI (INDIA)

Traditional Forestry Management

ISBN 978-93-5111-611-0

Published in 2015 in India by

RANDOM PUBLICATIONS

4376-A/4B, Gali Murari Lal, Ansari Road
New Delhi-110 002
Phone : +9111-43580356, 011-23289044, 011-43142548
e-mail: sales@randompublications.com,
info@randompublications.com, randomexports@gmail.com

Reprinted, 2018

Type Setting by : Friends Media, Delhi-110089
Printed at : Replika Press Pvt. Ltd.

Preface

Forest management is a branch of forestry concerned with the overall administrative, economic, legal and social aspects and with the essentially scientific and technical aspects, especially silviculture, protection, and forest regulation. This includes management for aesthetics, fish, recreation, urban values, water, wilderness, wildlife, wood products, forest genetic resources and other forest resource values. Management can be based on conservation, economics, or a mixture of the two. Techniques include timber extraction, planting and replanting of various species, cutting roads and pathways through forests, and preventing fire. There has been an increased public awareness of natural resource policy, including forest management. Public concern regarding forest management may have shifted from the extraction of timber for earning money for the economy, to the preservation of additional forest resources, including wildlife and old growth forest, protecting biodiversity, watershed management, and recreation. Increased environmental awareness may contribute to an increased public mistrust of forest management professionals. But it can also lead to greater understanding about what professionals do re forests for nature conservation and ecological services. Forest management varies in intensity from a leave alone, natural situation to a highly intensive regime with silvicultural interventions. Management is generally increased in intensity to achieve either economic criteria or ecological criteria.

I would like to thank my team for standing beside me throughout my career and writing this book. My special thanks go to "Random Publications" who have published the book.

– Ajay Mukherjee

Contents

1

Introduction

Forest management in the state is being carried out both in the traditional ways and according to the policies of the Forest Department.

THE TRADITIONAL REGIME

Traditional management systems have certain time-tested, practical and effective ways of managing as well as utilising natural resources. The Sarna (sacred groves) system, common in the northern districts of the State is an excellent example of sustainable management. The cutting of trees in these sacred groves is prohibited. Sanctified by belief and practice, this system has been an important factor in conserving the green cover of the State.

THE OFFICIAL REGIME

The Forest Department of the State manages the State's forest wealth in accordance with prescribed policies and guidelines. The Joint Forest Management (JFM) system encourages people's participation in managing forest resources. Members of JFM committees receive usufruct rights, a portion of the revenue from the felling of timber and from intermediate thinning.

They are also eligible for employment under afforestation and other programmes carried out by the Forest Department. The Forest Development Agency and the Chhattisgarh State Minor Forest Produce Co-operative Federation are involved in the management and development of forests and nationalised MFPs.

The Village Jan Rapats have suggested solutions for managing and maintaining forests. While these suggestions are quite varied, they do reflect a sense of disquiet at the denuding forest resources and the helplessness that people feel in the circumstances.

Other Issues Related to Forests

Other forest-related issues include the depletion of forests, their legal status and control.

Depletion of Forests

A common concern cited in many reports is the degradation and depletion of forests. The causes, according to the Village Reports, are the increasing biotic pressure on forests from the increase in human and animal population. Significantly, many Village Reports state that distancing people from the management of forests has also been a contributory factor.

According to them, since people have been alienated from utilising forest produce, they have become less concerned about conserving the forests. The people say unequivocally that they want to participate in preserving their forests. They also feel that unless they are fully involved in the work of protection, forests will continue to get degraded.

The Legal and Institutional Framework

The State Forest Policy guides the legal and institutional arrangement, based on the guidelines of the National Forest Policy, 1988, provided by the Central Government. Along with this, the provisions of Scheduled V areas in the Constitution of India and the Provisions of the Panchayats (Extension to the Scheduled Areas) Act, 1996 (PESA) also determine the legal situation.

The State Forest Policy asserts that the management of the forests should be such that forests are converted from an Open Access Resource to community controlled, prioritised, protected and managed resources through Joint Forest Management (JFM), People's Protected Areas (PPAs) and other such measures. The Government, through its Forest Policy, has made an attempt to recognise the ownership and relationship of the forest dwelling communities, especially the tribals.

The Policy states, "...For sustainable forest development, livelihood security and bio-cultural diversity conservation, People's Protected Areas (PPAs) should be established. This paradigm shift of adaptive management can reconcile the dichotomy of threat perception arising out of conservation-development orthodoxy by taking into account human sensitivities, socio-cultural norms, beliefs and systems borne out of history, culture and traditions."

From the tenor and content of the Jan Rapats, it is however apparent that the people and communities feel a sense of deprivation and alienation as well as a loss of access to their valued and valuable resource – the forests. The forest laws and its regulatory regime has divided the people from the forests and led to a realignment of the age-old relationship.

Village Jan Rapats repeatedly affirm that preserving and using the forests in a sustainable manner, was, and should be a way of life for the people. The forest laws and their implementation has resulted in the alienation of the people from their resources, and turned them into mere 'users' of the forests. Most people feel that the real control and therefore, the responsibility for the forests, now lies with the Government. They no longer feel a sense of ownership.

The Extension of PESA to Chhattisgarh

The Constitution provides for special provisions for administration and control of Scheduled Areas. The provisions of the Panchayats Extension to Scheduled Areas (PESA) Act, 1996, give special powers to the Gram Sabhas in Scheduled Areas especially in the management of natural resources. Areas with pockets of substantial Scheduled Tribe populations living within the dominance of non-tribal communities have been categorised in the Constitution as Scheduled V Areas.

Of the 16 districts in Chhattisgarh, seven districts (Surguja, Korea, Jashpur, Kanker, Bastar, Dakshin Bastar Dantewada and Korba) are categorised as Scheduled V Area districts and six (Bilaspur, Durg, Rajnandgaon, Raipur, Raigarh and Dhamtari) are partial Scheduled V Area districts. The objective of PESA is to enable tribal communities to safeguard their traditional rights over natural resources. The Act emphasises the rights and ownership of people's institutions and respects tradition in the control and management of resources.

It clearly states that, 'A State legislation on the Panchayats that may be made shall be in consonance with the customary law, social and religious practices and traditional management practices of community resources'. Further it states that 'A village shall ordinarily consist of a habitation or a group of habitations or hamlet or a group of hamlets comprising a community and managing its affairs in accordance with traditions and customs.' Some of the powers vested with the Gram Sabha in Schedule V Areas include:

- The ownership of minor forest produce.
- The power to prevent alienation of land in the Scheduled Areas and to take appropriate action to restore any unlawfully alienated land of a Scheduled Tribe.
- The power to exercise control over institutions and functionaries in all social sectors.
- Exercise control over local plans and the resources for such plans including tribal sub-plans.

Control, Ownership and Power Equations

Forests are a controlled natural resource. This control impacts substantially on the lives of people who depend on forests.

- For communities and households dependent on forests and for others for whom the forests sustain and supplement their livelihoods, accessing forest resources means contact with the Forest Department, the regulatory arm of the State. The unvarying threat of a powerful institution, with legal and physical resources to control this interface makes people, especially tribal communities, feel vulnerable and uncomfortable.
- Communities that have lived with the forests, managed and conserved

them for generations now find that the space for participating in forest management is dependent on the benevolence of the regulatory regime. They find this difficult to comprehend. The conservation effort is no longer natural but programme driven. People from all the villages state that the experience with officials and the mechanisms for interface are neither adequate, nor conducive to the common goals of society and State.

The issue of forest management involves a series of complex relationships between the stakeholders of the forests, the revenue department, the Panchayats and people. Regulations are perceived as being arbitrarily used by the Forest Department. This, combined with the restrictions imposed by the Government, causes friction between the people and the administration.

The critical balance between resource use and the issue of rights and people's ownership, and therefore responsibility of these resources especially in relation to forests, is an idea that the Forest Department is still coming to terms with. It is imperative for the State to define a role for itself vis-à-vis forests and the people who depend on them, in order to be able to stop forest depletion and encourage afforestation.

This will help to re-establish the vital balance in the forests of Chhattisgarh owned and managed for centuries by its people.

LAND

The land area of Chhattisgarh is about 1.35 lakh square kilometres. About 36 per cent of the area is cultivated, and another 44 per cent is under forests (forest land and revenue forests). Of the total land area in the State, 4,828 thousand hectares are sown, and the net sown area 13 per head is 0.24 hectares. The gross sown area is 5,327 thousand hectares.

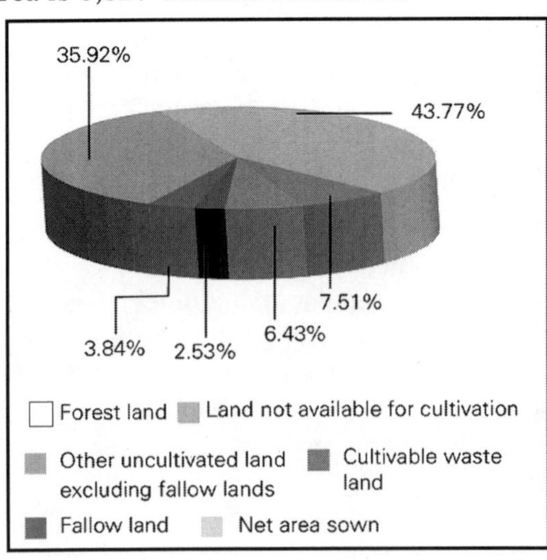

The highest percentage of land under agriculture is in Durg, Janjgir-Champa, Mahasamund (all above 50 per cent), followed by Raigarh, Bilaspur, Kabirdham, Rajnandgaon and Raipur (all above 40 per cent). The lowest percentage of net sown area to total area, is in Korea (18.7 per cent), followed by Dakshin Bastar Dantewada (19 per cent) and Bastar (21 per cent).

SOIL TYPES

Chhattisgarh has at least five different types of soil. In the districts of Bilaspur, Surguja, Durg, Raipur and Bastar red and yellow loamy soil is dominant. Both are low in nitrogen and humus content. A major part of paddy production comes from this region. In the hill ranges, the soil is sandy loam, which is also suitable for paddy. Laterite soil is good for cereal crops, while the black soil is best suited to cotton, wheat and gram. In the Jan Rapats, land has been categorised according to the traditional classification. This varies from district to district. The choice of the type of seeds, the crops that are sown and the technology that is used depends on this classification. It is a choice that has been tested and tried over generations and ensures some productivity irrespective of the quality of the land.

In many villages, the quality of land is not suitable for agriculture. While the undulating terrain and rocky surface is a constraint, the setting up of coal mines and coal related industries in districts like Korea, have meant that both land and water have been contaminated by pollutants such as fly ash. Many Village Reports have highlighted the fact that indiscriminate use of chemical fertilizers has affected land quality and led to a decline in productivity.

MINING IN CHHATTISGARH

Chhattisgarh is rich in mineral resources. Vast reserves of coal, iron ore and bauxite are found here, along with limestone and dolomite. This is the only State in the country where tin ore is found. Diamonds and semi precious stones like corundum, quartz and garnet are also mined here. While mining provides employment to some people and substantial revenue to the State, the industry has an adverse impact on the environment in some districts.

Pollution is one of the major impacts of mining, according to the Jan Rapats. This results in a number of problems ranging from declining productivity to contamination of drinking water.

- Village Jan Rapats mention that mining activities have affected the productivity of land and quality of water. District reports such as that of Korea have mentioned that the coal dust from coal handling plants covers the agricultural fields and affects the yield adversely.
- Pollution of the water that drains into reservoirs and rivers is another major problem. Several villages depend on surface water for domestic purposes, nistaari and irrigation. Polluted water has adversely affected both health and crops.

- Some reports have pointed out that illnesses related to breathing and respiration, falling levels of immunity, weakness and ill health are all outcomes of pollution. In some cases, people have been forced to migrate due the adverse impact of pollution on their health.
- Forest degradation due to mining activities has also been detailed in the Jan Rapats.

LAND DISTRIBUTION AND FRAGMENTATION

Land ownership and distribution are other important issues. The land distribution pattern is skewed, by the presence of a number of large farmers, due to benami land records and old malgujars (landlords who were earlier responsible for collecting rent on behalf of the State), who continue to operate in the central belt of the State. Increasing population and subdivision of holdings has led to tiny and unviable plots of land for small and marginal farmers.

LAND – AN ERODING RESOURCE

Soil degradation and soil erosion are increasing problems, leading to a decline in agricultural productivity. The Jan Rapats have noted this, and the following reasons have been cited:

- Pollution due to mining activities in the vicinity of agricultural fields and excessive use of chemical fertilizers.
- High cropping intensity without allowing the land to replenish the nutrient content and aerating the soil.
- Absence of good forest or vegetative cover, which leads to more soil erosion. The lack of good vegetative cover has also reduced dry leaves and twigs that fall on the land and which add to the productivity of the land.

LAND RECORDS

Two issues find frequent mention in the Jan Rapats:

- The problem of information on land records,
- Errors in the records.

Access to land records is not easy and the role of field level revenue officials is not always that of a facilitator. In many cases the records do not reflect actual ownership, especially in the case of larger landholdings. Another problem is that a large number of forest dwellers do not have clear land titles. Many of them have officially been categorised as 'encroachers' on forestland, although they were there long before the State declared their land as State forests.

ENCROACHMENTS AND DISPLACEMENTS

The issue of encroachments has been regularly cited in the Jan Rapats. While encroachments are present in almost all categories of land (private, State owned, open access and common lands), common lands have suffered the most, especially pasture and grazing lands. This has affected the quantity of fodder available for the cattle, especially for the landless, small and marginal farmers who depend on grazing and pasture land for feeding their cattle (they are unable to produce enough crop residues to feed their cattle).

Grazing is prohibited in forest areas and the continuously degrading forest cover does not provide enough fodder to feed their cattle, for the whole year. The Village and District Reports express serious concerns over encroachment. The political and power dynamics of these encroachments are such that people believe strong State intervention is essential for stopping and removing encroachment. There have been instances of families being displaced for the construction of dams, factories and industrial projects, and have not been fully rehabilitated.

FOREST MANAGEMENT ACTIVITIES

To facilitate efficient management, the forest has been divided into three blocks which cover areas of 91, 124 and 129 hectares. The intention is to protect and provide a sustainable and equitable utilization of the forest. Users have been involved in planting of various species assisted by forest technicians and committee members. In 2052 (1992), the women's groups planted 11,000 Sissoo seedlings. The following year the users planted 8536 seedlings of bamboo, Bakaina and Sissoo. There are plans to establish a plantation for Eucalyptus, Sissoo, Khair, Botlle Brush, Gul Mahar, Asoka and other species in the near future. The group is also undertaking NTFP farming in suitable areas under the guidance of forest technicians and other organizations. In collaboration with the Agro Herbal Company, a private company promoting NTFPs which recently set-up a field office in Dipyapuri, around 100 types of NTFP have been identified. The group has also established a tree nursery. The general level of awareness about community forestry has been strengthened and the popularity of the FUG increased. While people were afraid of DFO staff in the past, these relations have now improved.

PROTECTION

The protection wing has the overall responsibility for forest protection and has put in place rules and regulations to achieve this goal. Three user-managed protection posts have been set-up at strategic locations in the forest and three forest guards employed. Their job is to protect the forest against illegal felling, the use of fire and destruction of flora and fauna through grazing, poisoning, shooting and other harmful activities. Controls to regulate the collection of fuelwood, cutting of grass and pruning of fodder are in place and a graded penalty system where fines are calibrated according to the severity of violations is operative. There is a complete restriction on entrance into the forest from Chaitra 1st to end of Jestha as a precautionary measure to prevent forest fires. The group plans to construct a 5 m wide fire line in the north and west of the forest. Notice also that the forest is completely closed for grazing because of the high regeneration rate. The combined effect of these efforts is argued to effectively curtail illicit activities.

FOREST PRODUCT UTILISATION AND DISTRIBUTION

The Forest product distribution wing is responsible for the distribution of forest products and the process governing the allocation of timber is very similar to Dhuseri. Likewise, the forest product rates across the two groups are almost identical. Records are maintained for extracted and distributed forest products. To access timber quotas, users should submit an application along with ₹ 25 and state the reason behind the need for timber. After a recommendation (or rejection) of the application by the head of the sub committee, the committee will allocate timber as per the rules.

The role of the monitoring committee is to evaluate whether users have utilized forest products to fulfill their needs. The monitoring committee has a formal authority to penalise violations. It should be mentioned that while the mechanism for control of timber utilisation in Chautari is a replica of the process in Dhuseri, the above description provides the "official" account of the process of distribution of timber in Chautari. In contrast to Dhuseri, where we know that the control mechanism is ineffective, data limitations make it hard to establish precisely how well the mechanism operates in Chautari. While the incentive problems and scope for arbitrage are as strong as in Dhuseri, it is distinctly possible that the contrast between the official and the hidden economy is less pronounced in this case.

Branches and twigs damaged by the wind are distributed free of cost to users every year from 1st to end of Poush and in Jestha for 15 days but this is restricted to 2 people from each HH. The users can also collect the twigs and fallen small branches every Saturday during Shrawan and Bhadra. Users may also purchase fuelwood from the group for ₹ 100/quintal which exceeds the price of ₹ 75 charged by Dhuseri. Grass cutting is allowed from Bhadra 15th to Ashoj 15th and Mangsir 15th to end of Poush for the growth of the tree species. The forest is also open for fodder from Magh 1st to end of Baishak.

Special Provisions and Development Expenditure

Free membership has been granted to schools in the users' area of the CF. Timber is mainly provided to schools against a fixed price. The temple can get up to 5 cubic feet timber free of cost but have to pay for further requirements. The user group has awarded construction grants to Barchuli Junior High School Rajahar for extensions of classrooms and roof support to Saraswati Primary School in ward no. 4. A construction grant of ₹ 75.000 has also been given to the sub health post in Rajahar. Other financial and timber support has been provided to other schools, the police office, mothers groups and NGOs. Moreover, a grant for a biogas plant of ₹ 75.000, aimed at reducing fuelwood consumption has also been granted. The group has also given support in the form of disaster relief, mainly in connection with Jharahi floods.

Problems, Issues and Conflicts

Dhuseri and Chautari are undoubtedly advanced forestry user groups both in infrastructure development and in their respective approaches to community forestry. The user groups share another common feature: female leadership and participation remains very limited. Moreover, as noted, the motion for differentiation of users by socio-economic status was recently defeated in the General Assembly.

Based on a claim of being deceived by people living near the forest, *e.g.* residents in ward no 8, users from ward 5 have argued that a separate part of the forest should be allotted to them. Having much cultivated land and substantial livestock holdings, ward 5 has a high demand for forest resources to meet agriculture, domestic and livestock needs. The claim of deception was rooted in the observation that users in ward 8, adjacent to the forest, collect more forest products, legally and illegally. In 1999/2000 an interesting conflict emerged over a plan for a ward-wise division of the forest into plots. A demarcation for this purpose was undertaken on the initiative of the then Chairman X. However, users in ward 8, resisted this initiative. Having used the barren areas of the forest for grazing, a ward-wise division of the forest would effectively restrain their grazing opportunities. These users now accused the former Chairman, Mr X of being responsible for conflict claiming that he had received money from people from other wards after encouraging and backing their demands for a ward-wise division.

The committee pays field allowance of ₹ 12.-14.000 per year to DFO-staff during the utilization season. The role of the ranger during the harvesting period is to approve various steps in the harvesting process including estimates of felling, blazing and numbering of trees, to grant permissions for sawing in the sawmill as well as permissions to sell any surpluses outside the VDC. This provision of technical assistance falls within their official responsibilities and covered by regular salaries. Despite of this, the FUG is being charged for these services.

There is much disagreement about the payment and the scale of payment to the forest officials. The Nepal-German Ayurvedic society was prepared to enter into a 20-year agreement with the user group with a view to promote NTFP production and sales. An almost completed agreement was, however, undermined by strong opposition from the current chairperson Mr Y, and the proposal eventually scrapped.

His official claim was that the project could jeopardise the daily needs requirements of the users, due to a leasing clause in the proposed agreement. Mr X (the then chairperson) who had been supporting the initiative felt that this resistance threatened and undermined his leadership. There is a rather strong element of personal politics in this narrative. Mr Y's hidden agenda was to undermine Mr X's position to overtake the Chairmanship himself, a goal eventually accomplished, since the Ayurvedic society project compelled Mr X to resign. Mr X now claims that he can prove that the present chairman, Mr Y has been involved in illegal activities, more specifically, that Mr Y has felled green trees in conflict with the Operational Plan. However, Mr X is reluctant to provide further details because he is concerned about the reputation of the FUG.

During the chairmanship of Mr Z in 2057/58, an allegation of financial misconduct was raised in the assembly. It was found that he had spent ₹ 11000 to buy alcohol for the DF staff and for employing laborers for the transportation of woods from the forest to the depot. ₹ 11.000 was spent on food and travelling to the DFO in order to obtain the harvesting permit. A considerable sum of money was spent to persuade the ranger to approve a larger than previously agreed quota. Another case, accusing Mr Z for misuse of ₹ 9000 by giving favour to people already privileged through other FUG-activities, related to the loading of timber. The staff was given ₹ 9000 for loading in addition to their regular payments. Poor users argued that they should have been given the loading job and Mr Z was accused of denying the poor employment opportunities.

During his tenure, Mr X provided ₹ 1500/- to a DF staff in the harvesting season, *e.g.* December 2056/57 in connection with a visit to the District Office to obtain the blazing order for felling of trees. Mr X and the then vice chairman requested the ranger to produce the blazing order. The ranger lingered and asked for money. He was given ₹ 1000/- on the spot. The ranger then requested the chairman to buy lunch and incurred further expenses of ₹ 500. Two days later the blazing order was sent. After issuing the blazing order, DF staff marked 1700 cft for felling. While cutting it was discovered that some trees were hollow and the committee asked for permission to cut more trees to fill the allotted quota. The ranger accepted this request after extensive bargaining. Having been granted this second permission, the majority of the committee, who were aligned with the Chairman and with forest staff overstepped their mandate and felled old green trees as well. Inspecting the spot, the ranger gave the impression that the illegal felling had been reported to the district officer. He said that the committee should attend DFO for clarification without any further delays. The committee members were worried and asked the ranger to do whatever he could to minimize the offence. For this favour the ranger demanded 20 cubic feet of timber. The committee provided the timber immediately to save themselves from further trouble and embarassment.

Another interesting example concerns the sawing of timber. In July 2002, the ranger had given the committee permission to saw in Dibyapura Saw Mill at a rate of ₹ 38/cft.[3] Instead, the committee decided to process the timber at Pragatinagar saw mill (4 km west) which offered a rate of ₹ 30/cft, thereby permitting a saving of ₹ 8/cft. When informed the ranger sent a letter asking for clarification. He didn't approve of the answer thinking he had lost commission from the saw mill. The committee members involved in this incident are reluctant to provide further information about the matter.

FORESTS AND THEIR CHIEF PRODUCTS

In trying to see what lies ahead for world forestry it is first necessary to take cognizance of the dimensions and the complexities of the forest problem. It is also important to keep in mind the limitations of the forecasters, for the outlook depends very much on the lookout. In varying combinations, all men have hopes and fears, tendencies to look on the bright or the dark side, inclinations to hasty or slow judgement; some are dewy-eyed about the future and for that reason cannot see clearly, while others suffer from cataracts of vested prejudice that impair the vision. Furthermore, the lookouts are not all looking in the same direction.

Some can see only the truly remarkable achievements in technology and overlook the fact that for a long time to come the main wood problem of most of mankind will not be how to obtain rayon fabrics, cellophane wrappers, or molded monocoque airplane seats, but how to secure enough firewood for cooking purposes.

Others are attracted by the improved methods of silviculture, developed particularly in northwestern Europe, but fail to keep in view the hard fact that much of what can be done in Europe cannot, for economic reasons, be done in North America and probably cannot, because of different physical conditions, be done in the tropical forests.

This personal factor is no doubt the main reason for the greatly varying reports of the prospect in view, whether they deal with trends in agriculture, forestry, or any other aspect of the man-earth relationship. Some say that a wood famine is inevitable, others assure us that no world shortage of timber is in sight and the reason for the difference in opinion obviously lies in men, not in the forest.

THE PROBLEM DEFINED

Forecasting in this field is not an exact science, partly because of personal factors that cannot be eliminated and partly because the information on which the forecast must be based is neither complete nor altogether accurate. Forest mensuration is excellent in a few countries, but the available data on forest area, volume of stand, annual growth and even yearly consumption of wood in the world as a whole are at best only rough estimates. These estimates are nevertheless important. They are, in the first place, the only quantitative data we have and they reveal the dimensions of the problem before us.

The forest inventory published in 1954 by the Food and Agriculture Organization of the United Nations (FAO) shows that the annual world consumption of wood for all purposes in recent years has been over 50 billion cubic feet and that the total area of the forests on the earth is estimated to be about 9½ billion acres.

It is not easy, indeed perhaps not possible, for the human mind to comprehend such large quantities. Many have seen the forests of Washington and Oregon and been impressed by their vastness, yet in area they constitute less than half of one per cent of the world's forests. It is not easy to envisage 9½ billion acres of forest land under management on a sustainedyield basis, but that seems to be the future goal implied by the statement in Unasylva.

"The inventory shows that the world's forests are potentially capable of furnishing a plentiful flow of forest products for a world population much higher than that of today" It is not possible to say exactly how much forest land is managed on a sustained-yield basis at the present time, because data are lacking and because good practices have been initiated so recently in many regions that we do not yet know whether yield will actually be sustained. Management of a type is practiced on some of the forests in India, Ceylon, Burma, Australia, New Zealand, in other countries and in some parts of Africa. In 1949, according to the American Forestry Association, about 55 million acres of public and private forest land in the United States were classed as under "intensive management," which is defined by the words: "high order of cutting and good fire protection" As of January 1955 nearly 34 million acres of private forests in the United States were in certified tree farms.

However, not all of these forests are truly managed as the term is understood in northwestern Europe. It is perhaps safe to say that, after more than a century of research in silviculture and experiments with forestry practices, some 160 to 180 million acres of forest land are now under management in northwestern Europe and, with an optimistic view, possibly an equal amount outside of Europe, making a total of about 350 million acres. That is, it may be conjectured that 4 or 5 per cent of the world's forest lands are at the present time under reasonably good management and 95 per cent are not. The sheer size of the task that remains to be done if the goal is to be achieved looms very large in the outlook.

The task is not only large but complex. The ultimate goal might be defined as the utilization of the world's forest lands in such a manner that they will permanently yield sufficient wood for all mankind at a price people can pay, while at the same time some of the forests will provide watershed, soil and wildlife protection, grazing opportunity and facilities for study and recreation.

None of the terms in this definition can well be omitted. The term "forest land" must be used, for in the long run it is the land that constitutes the resource and not the generation of trees that happens to occupy the land at a given time. Forestry is only one type of land use.

In many regions it competes for site with other forms of land utilization: the demands of agriculture, for example, will have a decisive effect and the effect will be fatal to some of the forests. The term "permanently" must be included, for the problem is not merely to balance forest drain and growth in the immediate future, but to maintain the balance in perpetuity by increasing the growth to meet the rising demand of an increasing population. The phrase "all mankind" is required, for the goal cannot be regarded as having been reached so long as some forest lands produce species of trees that are considered useless because there is no market for them, while other regions are so short of wood that cow dung must be used for fuel.

"Price" is perhaps the most critical term. In one way or another and regardless of type of prevailing economic theory, every forestry operation and every forest product must be paid for, whether in energy, money, or barter. Good forestry practices are simply not possible unless they are economically feasible; for all we know, some of the forests may remain inaccessible forever because access may be too costly; and technological achievement, no matter how brilliant, will be of little use if people cannot pay for the products. The terms in the last part of the definition pertain to the concept of multiple use, an idea that may or may not turn out to be practicable in the long run.

In many parts of the world problems of soil and water are far more critical than those of timber supply and large sections of the forest, probably much larger than we commonly think, will have to be reserved for the primary function of watershed protection and will serve only secondarily, or perhaps not at all, as a source of wood. Other forest lands are withdrawn from ordinary economic utilization and reserved for recreation, study, wildlife sanctuaries, or for the purpose of saving some of the undisturbed forest for future generations. An example is found in the National Parks of the United States, which contain almost half as much timber land as there is in all of West Germany.

We may not always be able to afford this luxury, but as long as the world can keep its parks and wildlife reserves they will add up to a significantly large area which, like the protection forests, cannot be included in the area of the wood-yielding forest land. There are other complicating factors. The difficulties of silviculture and management are multiplied by the many different physical characteristics of the world's forests; people have different and deeply rooted attitudes towards trees; and the problem is not made easier by the fact that the forests are under about one hundred national jurisdictions.

It is not a mere matter of balancing numbers. Unasylva suggests that the wood needed for a moderately good standard of living would be an annual per capita consumption of about 35 cubic feet, basing the suggestion on prewar European experience; and makes the assumption, which seems plausible enough, that with reasonably good management the forests of the world can produce an annual yield of 30 cubic feet per average acre.

But it would be very misleading to conclude that the balance is favorable, or that no particular problem exists because there are only 2½ billion people in the world and 9½ billion acres of forest, or almost 4 acres per person.

Expected Demands for Forest Products in 1975

The outlook for the distant future depends largely on what will happen in the near future, which may be defined conveniently as the next two decades, since 1975 is the common target year in several recently published forecasts. 1 Only the major conclusions of these reports need to be considered here.

The Paley Report (1952) expects that the cubic-foot requirement of wood for all purposes in the United States will increase by 17 per cent between 1950 and 1975 and the Stanford Report (1954) foresees an increase of 14 per cent between 1952 and 1975.

These forecasts are thus not very different from earlier estimates: for example, in 1948 the Forest Service looked forward to a rise in the over-all requirement of about 50 per cent between 1945 and the year 2020, or an increase of nearly 15 per cent for each 25 years of that period.

The reports are in substantial agreement in predicting a decline in the demand of wood for pilings, railroad ties, shingles and cooperage and a very marked decrease in the use of fuel wood, but they differ on some items.

The Paley Report expects that the need of pulpwood will increase by 50 per cent; the Stanford Report, by 60 per cent. The demand for peeler logs, mostly for plywood, will rise by 40 per cent according to the Paley study; by 90 per cent in the Stanford forecast. The most notable difference is in the expected board-foot requirement of saw timber for lumber, which is by far the bulkiest item in the over-all demand: the Paley Report foresees an increase of 10 per cent; the Stanford Report, one of only 3.4 per cent.

The principal reason for this difference is no doubt the fact that the Paley Report is based on the assumption of no significant change in price relationships, whereas the Stanford Report assumes that the price of lumber relative to competing materials will increase and that lumber will lose some of its markets.

The Stanford assumption is well worth noting. High wood prices, causing people to prefer other materials, might have the effect of alleviating to some extent the coming pressure on the forest.

A world pulp survey of the same type as the Paley and Stanford studies is reported to be in preparation by the FAO, but, to my knowledge, projections of the future world demand for all wood are not yet available. A partial world forecast, which does not include the countries in the Soviet zone, is made in the Paley Report. The average total output and requirement of industrial wood during the years from 1947 to 1949 are compared with the average prospective output and requirement during the decade of the 1970's. Fuel wood and charcoal are not included.

The conclusion is that the total output will probably increase by 4 per cent and the total requirement by 40 per cent. It is very likely that the demand will rise, but the 4 per cent increase in output may be an overestimate. According to Egon Glesinger of the FAO, as reported in American Forests, the inventories have shown that the output of the world's forests is not rising, despite increasing demands.

The type of forest products demanded and regional differences between the United States and Latin America in the use of wood. The products are shown as percentages of the total cut of all timber, as measured in cubic feet.

	United States	**Latin America**	
	1952	**1975**	**1952**
Lumber	58%	59%	9%
Pulpwood	17	24	2
Fuel wood	13	3	80
Other products	12	14	9
	100	100	100

An indication of regional differences in the demand and use of wood on a world-wide basis is presented in Unasylva, December 1954. Expressed in approximate percentages of world forest area, population and total production in cubic feet of roundwood in 1953, the data reveal the differences between the highly industrialized countries of the northern hemisphere and the rest of the world. By "North America" is meant Canada and the United States and "Europe" includes the Soviet Union.

	Forest Area	**Population**	**Total Wood**	**Fuel Wood**	**Industrial Wood**
North America and Europe	40%	30%	70%	50%	88%
Rest of the world	60%	70%	30%	50%	12%

The heaviest production and consumption of industrial wood products, such as lumber, plywood, pulp and paper, are clearly concentrated in North America and Europe and the demand for these products is steadily rising. Furthermore, a large part of the industrial timber cut elsewhere, chiefly in the tropical forests, is exported to Europe and North America. The industrial demand is preponderantly for lumber and pulp, but the most widespread and still the largest single requirement is fuel wood.

Almost half of the total output of some 50 billion cubic feet in 1953 was fuel wood; even in Europe this requirement accounted for slightly more than one third of the total cut; and in many countries besides Latin America 80 per cent or more of the entire production was for firewood. At the present time, then, the demand on the world's forests is mainly for fuel wood, lumber and pulp, in that order and this is likely to continue at least in the near future.

Whether or not the forecasts will come true naturally depends on how valid the underlying premises turn out to be. The Stanford Report is based on the general assumptions that:

- No major war will occur.
- No radical advance in technology will increase production at a Tate faster than in the past,
- Business cycles will become more stable, accompanied by high employment. Another assumption, based on data of the Bureau of Census, is that the population of the United States will increase by 35 per cent to reach 212 million in 1975. The premises of the Paley Report are generally similar; for example, the expectation of a 40 per cent rise in the world requirement of wood rests on the assumptions of a population increase of 47 per cent, continued high demand in the industrial countries and rising consumption in others.

The safest of the assumptions is that population will increase and this alone would mean a rising demand for forest products. If a major war should come, the drain on the forest would surely increase drastically, as it has in past ways and the over-all requirement might become much larger than the forecasts indicate. If peace lasts but good economic conditions do not, the demand for industrial forest products would probably be less, but the world requirement of fuel wood might remain unaffected.

The second assumption of the Stanford Report has already been questioned by some forestry people, who believe that the large sums of money currently invested in research will result in technological improvements which will have a marked effect not only on efficiency in production but will help the industry to hold and even expand its markets.

An editor of House Beautiful magazine, speaking before the annual convention of the National Lumber Manufacturers Association in November 1954, challenged some of the Stanford predictions, declaring that the concept of abundant living demanded more lumber for dwelling units and urged the manufacturers to gear their thinking to an expanding market and an expanding American home.

It may well be that a 15 per cent increase in the over-all requirement of wood is about what can be expected in the next quarter century. But thereafter will come other quarter centuries and no evidence is at hand showing that this or any other country intends to stop growing in 1975. On the contrary, like the editor who spoke to the lumber manufacturers, most people hope for a steadily expanding economy, which of necessity means a steadily increasing demand on the natural resources, including the forests. If the hope is fulfilled, the outlook is that before very long the requirement will increase not by a mere 15 per cent, but by 100 or 200 per cent and perhaps more.

The question is how the forest will stand up under the rising pressure. The forecasters do not provide clear answers, only clear affirmations of their faith that the demand can be met. The conclusion of most of the reports, whether they deal with the United States or the world, is that the area of the forest is large enough to provide the wood needed in the future, even if the need should become twice or three times as large as today, but only on two conditions: much of the now inaccessible forests must be opened to utilization and all forests must be placed under greatly improved sustained-yield management. These are not small conditions and in them, it might be said, lies the whole forest problem.

The problem can be resolved into three major parts. The first is the question of supply: of forest area, forest soil, stand, annual growth and natural factors affecting the trees, such as fire, insects and diseases. The second is the problem of how best to utilize the supply: of silviculture, forest genetics, methods of cutting, logging, manufacturing-in short, technological knowledge. The third can be called cultural or institutional in nature; it is the question of our social-economic-political ability to apply the technological skill to best advantage.

More particularly, then, the outlook depends on trends of development in all of these fields; and the direction of the trends cannot be determined by looking only towards the future: we must also be guided by the past. Perhaps the most important thing to grasp is that in nearly all of these fields we are dealing with processes that have the tremendous momentum of history behind them.

SOCIAL VALUES IN COMMERCIAL FOREST MANAGEMENT

Apart from the policy, legal and ethical reasons noted above, there are good commercial reasons to consider social values. Forest managers are increasingly subject to pressures from other interest groups, frequently concerning social values.

- Demands from forest peoples' groups may include greater respect for local populations' rights, carrying out or desisting from specific management practices, and support for their own environmental/social projects.
- Demands from local groups to contribute to social and economic development may include support for local enterprise, employment opportunities, excision of specific areas from management, and use of company infrastructure.
- Demands from trades unions and their initiatives may include greater respect for workers' rights to organize and negotiate, fair wages and

benefits, health and safety at work, and the right to skills development throughout the organization.

- Demands to recognize other forest actors' rights to monitor, control and negotiate may include the development of agreements, and procedures to manage conflicts and make compensation.

Unless an active, organized approach is taken to respond to such pressures, forest managers may find themselves facing:

- slow-downs, strikes, blockades, boycotts, legal battles;
- damage to forest stock;
- arson, sabotage or vandalism to equipment and infrastructure;
- reappropriation of forest lands for cultivation, or migration;
- development of 'cultures of resistance' among forest-dependent people and their supporters (such as some consumers).

Considerable management skill and time may need to be invested in dealing with disputes, legal challenges and compensation claims, stalled negotiations, 'bad press' and political backlashes and general hostility towards forest managers and companies. In contrast, participatory approaches that develop people's potential contributions can benefit forest managers, by:

- improving the reputation of forest managers and companies;
- uncovering and sharing useful local information;
- broadening the base of ideas, skills and inputs applied to forestry;
- efficient apportioning of responsibility, *e.g.* local groups may be better suited to managing recreation and harvesting of non-timber forest products;
- better understanding of broader social, and thereby market, trends;
- improving transparency, accountability and therefore trust between all parties;
- longer-run cost savings and risk reduction.

By taking an active approach to people as well as to trees, forest managers greatly increase the potential social benefits from forest management and their chances of support from others. Such experience should also increase their capacity to anticipate future developments regarding social values, for example in legislation or the market, and thus to gain 'first-mover' advantage from the situation.

If some social values are indicators of market trends (and they frequently are), then it behoves forestry organizations to keep close track of them. Many companies have made good business ventures by diversifying into recreation provision in particular. And it is increasingly clear that there is considerable financial value attached to brand names of certain leading companies which are known to produce social and environmental benefits alongside fibre. For example, the Greenpeace name has been estimated to be worth hundreds of millions of dollars.

CODES OF PRACTICE AND CERTIFICATION STANDARDS ON SOCIAL ISSUES

All initiatives to define SFM principles and criteria cover social values. Some of these are becoming enshrined in legislation, while others face forest enterprises through market relations.

Current systems of forest certification specify various social requirements at the level of the forest management unit. Even so, social issues remain perhaps the most contentious of certification standards, and there are differences between standards, particularly regarding the required degree of involvement of different interest groups in forestry and treatment of peoples' rights. There are also differences in interpretation of the standards by certifiers/assessors. Whilst some confine themselves to assessing the local social outcomes of forest management, others assess the roles of local stakeholders in the management of the forest and indeed the enterprise, and call for changes that appear to reflect their own biases. Whilst it is increasingly clear that forestry should not be a principal means of social engineering, and less still should forest certification, it is generally agreed that social standards for forestry will become more stringent in future. Note that some require precise performance thresholds to be met, while others require only that the issue shall be measured. Still others consider that the social issue is a policy concern and require it to be covered in policies only.

ECONOMY OF FOREST MANAGEMENT

Forests play many roles in the development of a country, and especially in securing the livelihoods of people who live in and around them. Forest ecosystems are one of the greatest sources of biodiversity, but they are more fragile than many know. In particular, the natural forests of South and Southeast Asia, Africa, and Latin America are rapidly vanishing.

Although the international community has issued policy responses for sustainable forest management, forest degradation has not been halted in most developing countries. This situation requires a comprehensive analysis of the political economy of forest governance and an examination of the underlying causes of deforestation. Pakistan's forestry sector serves as an interesting case study for such an analysis in the South Asian context.

Deforestation in Pakistan is one of the highest in the world, despite rigorous institutional changes in forest management paradigms. This chapter attempts to provide an exploratory analysis of forest governance and deforestation and its consequences in Pakistan, to examine the interaction between forests and local livelihoods, and to identify the factors responsible for deforestation and the ineffectiveness of state forest management strategies. The paper argues that some of the main barriers to effective and sustainable forest management are a lack of understanding of local livelihood strategies, lack of political will on the part of state actors, lack of a sense of ownership of forests by the local communities, and the presence of powerful timber smugglers.

GLOBAL CONTEXT

Over the past few decades, the international community has discussed the global problem of deforestation and forest policy issues. In 1992, the United Nations Conference on Environment and Development in Rio de Janeiro served to catalyze debate and develop a vision of sustainable forest management. It is widely recognized that the forestry sector carries potential for achieving many of the Millennium Development Goals for poverty reduction.

The World Summit on Sustainable Development, the Kyoto Protocol, and the Intergovernmental Panel on Climate Change all recognize that forests are imperative to achieving overall sustainable development, reducing poverty, improving the environment, compensating for general biodiversity loss, mitigating the impacts of climate change, and ensuring food security. Despite these positive developments and a policy climate that advocates sustainable forest management at global, national and local levels, deforestation continues.

Development practitioners, donors, and policymakers must keep working to find sustainable solutions. Apprehensions about forest degradation and deforestation in many countries and regions throughout the world have given rise to numerous research studies about its causes and effects. There is a growing realization that unsustainable forest management strategies and insecure and conflicting land tenure and property rights are some of the main underlying problems of forest degradation.

Deforestation is one of the most significant global environmental problems. Patterns of forest degradation are particularly visible in many parts of Asia and Africa. According to the Food and Agriculture Organization of the United Nations, some South and Southeast Asian countries, including Cambodia, Indonesia, Nepal, Pakistan, the Philippines, and Sri Lanka are losing forests at rates exceeding 1.4 per cent per year. These are among the highest rates of forest loss in the world.

Within South Asia, the rate of forest depletion is highest in Pakistan, despite intensive support from international donor agencies and numerous global and local initiatives for forest conservation, policy formulation, and improved governance. In fact, most of the national governments of South Asia have launched major initiatives since the 1980s to decrease deforestation through structural reforms in the forestry sector, decentralization of governance, and community forestry initiatives, with a similar lack of success. Deforestation always brings negative consequences. In September 1992, Pakistan experienced the worst floods in the country's history, and the vanished forests in the northern watersheds were regarded as one of the main possible causes. Therefore, the federal government imposed a complete ban on logging in 1993.

But the ban did not take into account the country's own timber needs, and the ban not only triggered illegal logging there, but also led to smuggling of timber from Afghanistan into Pakistan, causing extensive deforestation in Afghanistan.

On October 8, 2005, Pakistan suffered its worst disaster in history, when an earthquake of 7.6 on the Richter scale struck South Asia, causing enormous destruction in the mountainous areas of northwest Pakistan. Massive landslides caused further loss to the region's inhabitants. The landslides occurred mostly in the denuded hills, whereas places with good forest cover suffered less destruction.

FORESTS OF OTHER COUNTRIES

According to statistics issued in 2006 by the Ministry of Finance and Economic Affairs, forests cover about 4.22 million hectares in Pakistan, only 4.8 per cent of the total land area. However, there is considerable controversy over the precise forest area in Pakistan, as different national and international agencies have published statistics based on different definitions of what comprises a forest. Areas designated as "forest areas" are merely lands under the administrative control of the provincial forest department.

Thus, officially designated forests may be devoid of trees while considerable tree cover may be found in areas other than the designated forests. There is a large variety of tree species because of the country's diverse physical geography and climatic contrasts. The important forest types are hill coniferous forests (46 per cent of the total forests), scrub or foot hill forests (28 per cent), irrigated plantations, farmland trees, and mangroves in the delta of the Indus River. Most of the forests are found in the northern part of the country, with 40 per cent in the Northwest Frontier Province (NWFP), 15.8 per cent in northern areas, and 6 per cent in Azad Kashmir. Eighty per cent of the forests in Pakistan are naturally distributed in the Himalayan, Karakoram, and Hindu Kush mountain ranges. Although Pakistan's forest resources are scarce, they contribute significantly to its economy. These forests are imperative for the protection of the natural environment, production of various goods and services (such as timber, firewood, and medicinal plants), and the protection of land and water resources, particularly in prolonging the lives of dams, reservoirs, and the irrigation network of canals in the lowlands, where intensive agriculture is practiced.

Legal Classification of Forests

The provincial forest departments are charged with governing the forests, while the federal government is mainly responsible for policy formulation and international matters. The natural forests are managed according to their legal classification and tenure rather than according to species. These forests are divided between state and non-state forests. More than two-thirds of the total forests are state-owned and are generally divided into reserved forests and protected forests. In reserved forests, the local people have very limited rights. They are only allowed to collect wood for fuel and extract timber for their personal needs. The main category of non-state forests is the subsistence (*guzara*) forests in which the owners or holders of exclusive rights are entitled to use the forest wood for domestic purposes. Others may be given permission by the owners for certain uses, such as grazing animals and collecting firewood. Provincial forest departments are responsible for management and planning of all types of state- and non-state-owned forests, except farm forest areas.

Forest Tenure as a Source of Conflict

There is a wide gap between the legal status of forests and the actual practice of forest management. In some areas, state control of the forests is never accepted by the locals, particularly in those forests where traditional rights have long been recognized. In some cases, local communities still claim ownership of these lands. An especially interesting case is that of the protected state forests. Based on traditional institutions such as customary land titles, many local people are of the opinion that they themselves own the forest. They do not accept legal ownership by the state, even as state authorities strive to assert their legally designated control. Such conflicting interests between the state and local communities have placed forests under continuous strain. Uncertainties and inequalities regarding tenure are a major cause of forest depletion. The local communities perceive the state to be in competition with their interests rather than being a mandated caretaker of the forests. Recent empirical studies have indicated a marked communication gap and distrust between the state and local stakeholders.

Illegal Logging

Timber harvesting from the mountain forests of northwest Pakistan has been banned since 1993, following the destructive floods of 1992, but illegal logging continued after the ban because of high demand for timber in the cities. Timber prices in Pakistan escalated after the ban, making illegal timber harvesting and smuggling from the highlands to the lowlands a very profitable business. The term "timber mafia," which came into common use after the ban, refers to a network of timber dealers, corrupt politicians, officials of the forest department, influential tribal leaders, and others who make money by illegally harvesting and smuggling trees from the highlands to the lowland cities. They rely on bribing, bullying, political networking, and blackmailing. Powerful politicians, including members of Parliament, are believed to support or be part of the timber mafia. It is widely believed that these individuals can manipulate legislation to serve their interests and resist changes in forest law that would make forest management more participatory and sustainable. The civil society and media in Pakistan often accuse the forest department of being involved in illegal logging.

2

Benefits of Forest Management

Apart from firewood and fodder, which are direct products, considerable income is derived from the forest from the sale of timber, from the collection fees, from eco-tourism, and from funds from other organisations. For example, in 2006 Baghmara BZCF was awarded the prestigious King Gyanendra Nature Conservation Award, with prize money of 100,000 rupees, by the Royal Nepal Academy of Science and Technology (RONAST), for contributing to sustainable development by promoting eco-tourism and conservation of biodiversity through community forest management. These funds are used to support a variety of community development projects. Many of these are of a general nature and in principle benefit the village as a whole (road improvement, embankments, schools etc.), but others are targeted towards individuals, in particular the projects for training in income generation activities. These include bee-keeping, seasonal vegetable farming and animal farming. In addition, financial support is given to individual families for construction of toilets, rice husk stoves and biogas plants, in the form of loans.

These benefits do not reach all families equally. The Musahar women mentioned that they have not received any kind of training, only few are enrolled in adult literacy classes. In any case they do not have sufficient money to start any micro enterprise and cannot raise animals as they do not have land. So although the programmes devised by the executive committee are intended for poor and marginalised women, they are often in practice of little relevance to them. Most of the training sessions and workshops are in fact attended either by the wealthy or the middle class groups. "Weaker groups" are unable to attend as they are day labourers, and their families will go hungry if they miss a day's work (the workshops generally provide a meal for the participants, but the families of these participants of course do not get fed). One woman member of the executive committee explained that they try hard to bring poor and landless people into income generation training but they do not come. Most marginalised people, the poor and particularly poor women indeed leave their houses early in the morning to work as labourers in the road or building construction industry in the city and return home only after dark.

As regards the issuing of loans for the purchase of equipment, particularly for biogas, the "weaker groups" say that they do not benefit at all. The research showed that biogas is mostly installed in wealthy and middle-class houses, which is not surprising as the loan only covers part of the total cost, and only these families are able to pay the extra money needed for the installation. Moreover, it is only the wealthy and middle class that have enough cattle to supply dung for a biogas plant, and can afford to stall-feed them close to the house, which is necessary for transferring the dung to the biogas plant. The poor have fewer (or no) cattle, and lack the space to build stalls close to their houses, and the time to gather fodder for stall feeding. The poor do not take loans for other equipment such as toilets and husk stoves because they do not have any collateral and in any case they often have difficulty paying back the interest.

From this one can conclude that distribution of the benefits of the community forest management effort are not equally distributed within the community. It is not necessarily the case that this mal-distribution is deliberate on the part of the FUG and its executive committee, although the exclusion of the Musahar people does seem to indicate on-going bias. It is more that there is deep-rooted, structural inequality within the village already, which is very difficult to overcome. Indeed it would be very surprising if a single programme like community forest management were able to totally change these economic and social relationships, although recognition of the problems, and efforts to design community forest management procedures which take them better into account, could certainly be improved.

BENEFITS OF FORESTS

Trees are necessary for our survival. Through photosynthesis trees produce the gas that we cannot live without: oxygen (O_2). As we breathe in, our bodies take in oxygen and when we breathe out, we release carbon dioxide (CO_2). Trees do the opposite. They take in CO_2 and release O_2. This cleans the air by removing harmful CO_2 so that people and animals can breathe.

A tree is often the symbol used to represent the environmental movement. In fact, we adopted the colour green, the plant's colour, as the main adjective to define general environmentalism. Why is the forest such an important element for the environment? There are many reasons why we benefit from forests and there are obviously many consequences that stem from the loss of 13 million hectares of forest every year.

Forests stabilize the climate in general. The plants enrich the soil by recycling the nutrients through the shedding of leaves and seeds. They also regulate the water cycle by absorbing and redistributing rainwater quite equally to every species living within its range, which is known as the economy of water. Overall, forests provide perfect habitats for life to flourish on land. They actually contain most of the living species, particularly in the case of tropical forests where up to 90 per cent of the planet's species live. Tropical forests possess the highest level of biodiversity and therefore provide the biggest genes reservoir.

LIVING OFF THE LAND

Earth without forests is a picture that most of humankind presently could not conceive. Forests cover much of the planet's land area. They are extremely important to humans and the natural world. For humans they have many aesthetic, recreational, economic, historical, cultural and religious values. Timber and other products of forests are important economically both locally and as exports.

They provide employment for those who harvest the wood or products of the living forest. Herbalists, rubber tappers, hunters and collectors of fungi, nuts, bamboo and berries are able to utilize such resources. Other non-wood forest products come in the form of medicinal compounds, dyes and fabrics. There are many people who are dependent on forestland for their livelihoods. One-third of the world's people depend on wood for fuel as a significant energy source. Surveys in Cameroon, Cote d'Ivoire, Ghana and Liberia found that forest wildlife accounted for 70 to 90 per cent of the total animal protein consumed. Some indigenous peoples are completely dependent on forests. As well as providing a home for some people, the forest environment provides a popular setting for eco-tourism, which includes hiking, camping, bird watching and other outdoor adventure or nature study activities.

Protection from Natural Disasters

Trees protect the soil against erosion, and reduce the risk for landslides and avalanches. They may increase the rate that rainwater recharges groundwater, as well as control the rate that water is released in watersheds. They help to sustain freshwater supplies and therefore are an important factor in the availability of water, one of life's basic needs. When rain falls, some may sink to the ground, some may run off the surface of the land, and flow towards the rivers and some may evaporate. Running water is a major cause of soil erosion. During heavy rains, flooding may occur, filling the waterways with eroded soil. The silt clogs these waterways, cutting off water sources for plants and animals during the dry season. Silt may also fill reservoirs created by dams, reducing its ability and future capacity to generate hydroelectricity and provide irrigation. The removal of forests causes nutrient loss in the soil especially if the period between harvest isn't long enough.

Scientists have recognized that trees can also serve as a tool in the reduction of storm water run-off. The incorporation of trees and other vegetation costs five to ten times less than using solely manmade storm water infrastructures. The leaves on trees keep large quantities of rain and snow from falling to the ground and tree roots absorb excess surface water, thereby stabilizing ground soil. Street trees provide the greatest annual benefit in avoiding storm water run-off by diverting 327 gallons of water compared with the 104 gallons averted by park trees.

Helping the Climate

Researchers have found that trees help the urban ecosystem by decreasing air temperatures. Studies indicate that a 10 per cent increase in tree canopy cover results in a one to two degree Fahrenheit reduction in air temperature. In addition a one-degree decrease in temperature will reduce the possibility of smog by 6 per cent. Furthermore, increased tree canopy coverage protects urban dwellers from harmful effects of ultraviolet (UV) radiation. Strategic planting of trees can also increase a city's energy efficiency. Research conducted since the mid 1980s has quantified the energy saving potential of urban forests. According to the Energy Information Administration, household heating and cooling cost consumers 180 billion US dollars in 1987. Studies have found that a 25 feet tall tree could save 10 to 25 US dollars annually on these energy costs alone. Because trees release cool vapour into the air during photosynthesis, the need for artificial cooling devices is reduced. In fact, according to one study, the air-conditioning savings from a deciduous tree near a well insulated home ranged from 10 to 15 per cent, while an 8 to 10 per cent savings was reported during peak cooling periods. Landscape vegetation around individual buildings can also result in heat savings of 5 to 15 per cent savings, and cooling savings of 10 to 50 per cent.

It has been observed that one of the largest energy fluxes at the earth's surface is that due to evaporation by trees. Heat is absorbed by trees for transpiration of fluid, and later released into the upper atmosphere. The fluid involved is groundwater flowing up the trunk of the tree. Increased annual run-offs from deforested areas in the Amazon support this attribution.

Purification of the Air

Forests affect the climate and are an important source of oxygen (O_2), although they play a lesser role than once thought.. Rain forests serve as an important filter for carbon dioxide (CO_2), a greenhouse gas that contributes to global warming. The Amazon region alone stores at least 75 billion tons of carbon (C) in its trees. When stripped of its trees, rainforest land soon become useless and inhospitable because the soil lacks the nutrients to support any kind of agriculture. Regeneration of a tropical rainforest may not be possible or, when it can occur, it may take hundreds of years. Research continually reveals that trees benefit urban communities in a number of ways. First with respect to air quality, trees remove damaging pollutants from the atmosphere, and replenish it with O_2. Through the process of transpiration and photosynthesis, trees sequester grams of ozone (O_3), sulphur dioxide (SO_2), nitrogen dioxide (NO_2), and carbon monoxide (CO) every hour, amassing several tons of carbon storage each year. This carbon sequestration process in turn reduces the harmful effects of these noxious gases that cause global warming as well as lung-related ailments. Researchers have also been able to quantify the value of this carbon removal through the use of a carbon storage and sequestration model called UFORE-C. In fact utilizing the figures economists employ to estimate the effect pollutants cost society, one research ecologist was able to compute carbon sequestration into a tangible "dollar-value.".

CAUSES OF DEFORESTATION

There are many causes of deforestation however harmless they may seem. So much damage can be done by even a single chainsaw, because behind those chainsaws are huge companies that care only about demand and profit, and forests are needed to supply this.

Commercial Exploitation

The first and most important cause of deforestation is wood extraction. Wood has always been a primary forest product for human populations and industrial interests. Since wood is an important structural component of any forest, its removal has immediate implications on forest health. Intensive harvests can lead to severe degradation, even beyond a forest's capacity to recover. When the soil has been stripped of its nutrients, farmers move further into the forests in search of new land.

Shifting cultivation is one of the most unproductive uses of farmland, and a major cause of degraded land where forests cannot regrow. In eastern India, this agricultural practice called "jhumming," has laid barren previously fertile tracts of the hillside. Timber is one of our most precious as well as used resource. We use it to build our houses, furniture and stock our fireplaces. This heavy demand fuels the destruction of our forests at an unsustainable rate. For every tree that is logged, 27 are killed or damaged in the process. In small doses logging isn't too bad, and over 30 or so years the forest will grow back. But with today's demand for wood the areas that are logged are too large, and this causes permanent destruction of the forests. The impact of the timber trade is generally greater than has been claimed in the past. The North plays a key role in many of the factors leading to forest decline.

Commercial forestry is the leading cause of deforestation in the world's temperate regions. The forces of large global markets for wood and wood products drive the scale of logging activities such as clearcutting. The source of demand is increased consumption by North America and Europe, not population growth. Again, transportation routes have a role, opening up new areas for natural resource exploitation. Privatization of natural resource industries has led to decreased regulations regarding timber harvesting. Multinational corporations dominate trade in wood. Most of these companies were organized in the US. Multinational companies for whom improvement of forest practices is not a priority often export the timber in an unprocessed state out of the country of origin.

Mining for precious resources also plays a major role. There are many forests that hold fair amounts of Earth's resources, such as iron ore, copper, oil and other precious metals. Many mining methods such as strip-mining and strong-force hoses break down the earth and cause major erosion. The mining sites are large and many trees need to be demolished to make way for them. When nothing is left to be mined, there is little chance of the forest growing back because of erosion and the lack of nutrients in the soil, which was churned out during mining.

Development of Degrade Forests

People destroy or degrade forests because, for them, the benefits seem to outweigh the costs. Underlying causes include such issues as poverty, unequal land ownership, women's status, education and to some extent, population. Immediate causes are often concerned with a search for land and resources, including both commercial timber and fuelwood. In many areas, rural households rely solely on fuelwood collected from the forest for their domestic energy supply. The roads that are built into the rainforest encourage and provide access for settling activities. In north-east India road building is often wrecking havoc on the forests. A road is cut through a hill face and the first loss is of the trees along its trace. The debris is thrown down, destroying the trees below, leaving a trail of dead or wilting trees in its wake. This debris enters the valleys, pushing up the levels of streams and rivers, causing siltation and floods.

There are many government agencies with policies that are uncoordinated in nature. Long range planning is not undertaken; and the Amazon is greatly affected by forces outside of the region. Some of the causes of migration to the tropical forests are population growth and political persecution. The settlers clearing and cultivating the land do not have the knowledge and experience of indigenous peoples of the forests and are unable to utilize the land effectively or sustainable. The process of shifting cultivation is accelerated and as a result the forest doesn't have enough time to recover. Tropical rainforests are truly under the assault by humans.

Cattle Ranching

Perhaps the worst culprit of deforestation, at least in the Brazilian Amazon, is cattle ranching, accounting for 38 per cent of deforestation in this region. Cattle ranching involves hundreds and thousands of cattle grazing on expansive areas in and near forests. Because the forest soil isn't adapted to these conditions, it isn't long before the area becomes unproductive. So cattle ranchers expand their grazing area, leading to more destruction. After the grazed land is left, the forest is very unlikely to grow back due to the stripping of the soil. Cattle ranching each year in this manner destroys an estimated 5700 square miles of rainforest alone. The Brazilian government subsidizes some of the cattle ranches that exist on converted forestland. The land is unproductive. Much of the demand for the beef comes from the fast food hamburger market, which is more concerned with quantity, than quality farm raised meat.

Agriculture, Ecosystem and Forests

Monocultural forestry simplifies the ecosystem, leaving it vulnerable to disease and other environmental factors. In the tropical forests of the world, the clearing of land for agriculture and livestock are the primary activities resulting in deforestation. The main cause is unequal distribution of land. 4.5 per cent of Brazil's landowners own 81 per cent of the country's farmland, and 70 per cent of the rural households are landless. It seems that these conditions cause people to encroach on, penetrate and modify the forests. Governments have an important role in these processes. In many countries in Asia and Africa, where family farms are still prevalent, the breakdown of large joint-families is causing uneconomic divisions to existing farms. These inefficiencies, in turn, put pressure on farmers to sell their land for development, and it turns whole farming comminutes into new developments.

Poverty and Inequality

Another cause of the ecological crisis of the present is social inequality. Gender inequality is one of the more powerful forces at work, which exists in virtually all of human cultures. The natural world is often portrayed as feminine, in terms such as "Mother Nature," "virgin forest," "exploitation" and "rape of the valley" that are used to describe elements and uses of nature and serve to perpetuate this harmful attitude. Human society's attitude with regard to the status of women makes an important contribution to environmental degradation and deforestation.

Although it is easy to assume a strong connection between population growth and deforestation, some research indicates that the problem is more complex. It involves non-demographic mechanisms resulting from credit and capital market failures, lack of suitable mediating institutions securing property rights, wretched poverty, uneven land distribution, consumption patterns in developed countries, greedy multinational companies, ignorance and bad management by colonist or frontier land, and so forth.

One of the underlying causes of human exploitation and consumption of forests and other natural resources is human tradition and beliefs. One source of such belief is Christianity, whose dominance in the Americas and Europe has important consequences for natural resources. Christian's attitudes are of anthropocentrism. The dominant power on the planet is humankind. The first human, Adam, gave the animals their names, shows this kind of dominance. Modern science and technological changes began in the name of Christianity. These beliefs created the attitudes, traditions and activities that enable us to be responsible for the destruction of nature that is occurring in the forests and the rest of the world. It is then possible that if certain Christian beliefs were different, human attitude towards nature would be that of conservation not exploitation. Such contrasting beliefs could include a god or gods that exist on Earth or even in trees, or that humans are reincarnated as plants and animals.

RIGHTS OVER FOREST RESOURCES

The tribal people are facing serious problems with regard to utilisation and rights over forest resources. Due to the increasing pressure on forests by various interest groups, there is a corresponding pressure on the tribals to reduce their dependency on forests. This is creating serious situations of conflict, as tribal life is symbiotic with land and forests and their livelihood and culture are based on their relationship with the natural wealth around them. The tribals are being harassed for using forestlands and being evicted in many places. Such reports have come from places like Khammam, Visakhapatnam, Vizianagaram, Adilabad, Srisailam and other places. In Khammam in one particular village, the forest and police departments allegedly branded the tribals on their shoulders as an indication that they were destroying the forests. There are ambiguities in forest-revenue land demarcation. In some places like Nellore district, a lot of land on Velugonda hills is indicated as 'poramboku' in revenue records and as RF as per the forest department. These lands do not have any forest growth. The FD is taking up palm oil plantation in these RF lands (which is not a forest species). However, landless tribals are booked in criminal cases or prohibited from using these lands for agricultural cultivation. These lands should be given to tribals with pattas for cultivation.

In some areas like in Visakhapatnam district, the lack of clear forest boundaries is making tribals vulnerable to the exploitation of both the forest and revenue departments. A joint survey and demarcation of boundaries by both departments should be immediately taken up to arrest these conflicts. Such Joint surveys need to involve the villagers at various levels. The Forest-Revenue Boundary dispute is a perpetual problem in Adilabad and Warangal districts, leading to booking of cases by the Forest department and tension in these tribal villages.

In Visakhapatnam Agency and Nallamala areas, there is the unique problem of Enclosure Villages. There were many tribal villages that were not enumerated in the forest surveys. Due to such sheer negligence, the villages were not given revenue status and to this day, they do not have pattas for their lands. They face constant harassment from local forest officials, as they do not possess land records. Recognizing these enclosure villages and issuing pattas to tribals should immediately resolve this problem. In Buttapur (Adilabad district) and in Nellore Dt. (Yanadis), the tribals were given lands decades ago under the social forestry scheme and are cultivating there. But due to lack of pattas, they are being harassed by the police and forest departments and also do not have access to bank loans as they cannot prove their ownership. These tribals have to be given pattas as promised.

Attacks on tribals, their properties and livestock by wildlife are not compensated by the Forest Department. Several cases are pending where tribals have been either killed or disabled and yet have not received any monetary compensation as due to them under the Wildlife Act. The tribals should not be prohibited from entering the forests to collect NTFP for their domestic requirements, like firewood, medicinal herbs, food, or agricultural and housing material.

The tribals should not be defined as 'encroachers', as is being projected by the forest department and also as indicated in the Circular of IG Forests, MoEF dated 3rd May 2002. In districts like Visakhapatnam, East Godavari and Vizianagaram, where there is high prevalence of podu cultivation, the tribals are facing threats of eviction from the forest department. There was a notification issued by the government of A.P in 1987 ordering that pre-1980 settlements will not be evicted until further orders. This should be implemented. The JFM programme now renamed as the Community Forest Management programme of the A.P Forest Department has caused grievous violations with regard to tribal rights. One major violation is the displacement of tribals from their podu lands by reclaiming them back into the forests through the JFM programme. The official reports of the forest department and the World Bank (which has funded the project) reveal that 37,000 hectares of forestland has been reclaimed back from the people§. The APFD further states its intention of displacing tribals in G.O. where it calls for voluntary surrender of lands by involving NGOs to motivate people under the CFM programme. The APFD has come up with a proposal for rehabilitation of tribals displaced from their podu lands through a monetary and schematic approach. This should be condemned as it bypasses the issue of eviction and rights of people over their podu lands. Their act of reclaiming lands under JFM programme and the APFD's forestry project under World Bank assistance has to be scrutinized closely by the Commission.

Samata and a few other NGOs raised this issue by writing to the World Bank on its violations of its own Operational Directives 4.2 and 4.3 in its financial support to the APFD. This has led to the Bank's insistence on an R&R policy to be proposed and implemented by the APFD in its CFM programme In Srisailam area, the Rajiv Gandhi Tiger Sanctuary has led to eviction of tribals from their original homes. They have not been properly settled so far. The concept of EDC (Eco Development Committees) that was introduced is working to the detriment of the tribals, as the objective of this programme was to reduce the forest dependency of the tribals. The tribals are given income-generating programmes and alternate sources of firewood by the forest department. This is an artificial mechanism that is not sustainable in the long term.

In the Srisailam Tiger Sanctuary area, the Chenchus, who are traditional hunter-gatherers, go into the forest everyday for all their needs. They are being harassed by the forest department for trespassing into the sanctuary. The groups who came from the chenchu area felt that the Chenchus should atleast be given identity cards to prevent harassment from forest department and the police department who mistake them for naxalites.

Another serious problem with regard to Sanctuaries is the settlement of people's rights. The Settlement Officer is the Conservator of Forests whereas the forest department is one of the interested parties in the land acquisition for the sanctuary and hence it is not appropriate to make the forest officer the settlement officer**. It should be the revenue department (District Collector) who should be authorised to settle the people's rights. G.O.No.112 which calls for involvement of private industries like ITC, Reliance and others for taking up commercial plantations through the VSS should be withdrawn immediately, as this is a backdoor method of allowing private industries to enter forest and tribal lands. Such tripartite agreements between the government, the industries and the tribals can never provide a level playing ground for the tribal people and make them more vulnerable to the exploitation of private industries.

The historical injustices to the tribal people in this region and that of the tribals in southern Orissa due to construction of several 'development' projects and industries like Nalco, HAL, Sileru, Machkund and others which have displaced tribals in large numbers without any rehabilitation, and has forced them to migrate in search of livelihood. They are now being treated as encroachers and criminals in forestlands. This definition should be condemned and Commission should recommend proper mechanisms for resolution of conflicts. The Commission should strongly recommend the orders passed by the Ministry of Environment and Forests in 1990, on the recommendations of the Commissioner of Scheduled Castes and Scheduled Tribes, to ensure that such conflicts are resolved. Infact, the first two orders alone show a way forward to settle disputes concerning forestlands.

All the groups consulted strongly felt that the pressure on forests and the destruction of forests is taking place more due to increase in industries and private commercial activities like Mining, paper-mills, timber smuggling and the growing non-tribal population settling down in Agency areas, than due to the podu cultivation practised by tribals. Eco-Tourism is also causing destruction of forests. For example, the A.P Tourism Department built huge infrastructure close to Farhanbad, near Mannanur (Srisailam) for Tourism purposes in the middle of the protected area and in Araku, Borra and Anantagiri (Visakhapatnam). While tribals are prohibited from entering these forests for their survival needs, it is totally unjust to allow tourism, which is causing a lot of degradation. Hence, the constitutional provisions of the Fifth Schedule, the Samatha Judgement and the PESA Act should be strictly implemented following the right spirit with regard to industries and non-tribal settlers, which includes the Tourism Department.

The Tourism Department also conducts tours through tribal villages, especially in Chenchu settlements, and exhibits them like museum pieces. The groups felt that this degrading exercise further underscored the lack of empathy among the various government departments towards the tribals.

Because of a shift in the policy of the government, profits to G.C.C. are more important than benefits to tribals. Although there are 32 items in the procurement list of GCC Ltd, it procures a limited number of commodities and at very low prices compared to the market rates. The tribals face several constraints in collection of NTFP and marketing these items because of the monopoly rights enjoyed by GCC Ltd. They do not have the right to sell their produce to private traders even when their rates are higher. Hence, a long-standing demand has been for removal of the monopoly restriction of GCC Ltd. and instead a minimum support price should be provided by GCC to protect the tribals from exploitation of traders. The monopoly of forest department over Beedi leaf collection and sale should be removed. The tribals or VSS should be given 100 per cent rights over income from beedi leaf and also the right of sale, with the forest department providing the support price.

INDIAN FOREST LAW AND POLICY

After Independence, however, and according to the newly enacted Indian Constitution, forests were placed under the so-called 'state list' according to which state legislatures have a primary right to make laws. In 1976, the Indian Forest Act was added to the concurrent list of the Constitution of India, giving the Central Government and states shared responsibility and control over forest matters. Thereby the Government of India does have the power to legislate on forestry issues but only after consulting the states. The responsibility for administering the forests lies primarily with the state government. The relative position of the Government of India depends on several factors, such as its contribution to the state budget and state polities. The balance of power between central and state governments has remained a key issue in forest management ever since.

After 1947, in a post-independent India, commercial exploitation and degradation of India's forests increased dramatically. Indeed, the 1952 National Forest Policy set out guidelines which were, for the most part, directed towards the supply of cheap timber and non-timber forest products for state-sponsored industrialization and modernization. In 1980, with the passage of the Forest Conservation Act, the central government reasserted some of its control over forest-based resources. The Act restricts the state government's power to de-reserve a forest, and it restricts the use of forestland for non-forestry purposes without the prior approval of the central government. Unilateral decisions from the centre have prevailed from then onwards. Only six centres were set up to monitor forests and conservation (as per the Act), which is obviously insufficient for effective regional implementation in a country the size of India. Thus the Forest Conservation Act of 1980 has been problematic for a number of reasons, and has achieved little improvement in the 'conservation' of India's forests.

The National Forest Policy of 1988 envisaged people's involvement in the development and protection of forests *for the first time*. It stipulated that the requirements of people living in and near forests for fuel wood, fodder and small timber should be treated as the top priority, and forest communities should be motivated to identify themselves with the development and protection of forests from which they derive benefits. A primary task of all agencies responsible for forest management, including the forest development corporations, should be to engage tribal peoples closely in the protection, regeneration and development of forests, as well as to provide gainful employment to people living in and around the forests. However the 1988 National Forest Policy has never been translated into law. It remains essentially a broad statement of government intent and does little in the way of specifying any legal rights or duties owed to forest communities-especially the tribal/indigenous peoples.

NATIONAL FOREST POLICY OF JOINT FOREST MANAGEMENT

One outcome of the National Forest Policy was the launching of Joint Forest Management (JFM) initiatives by all the States and Union Territories. However, there have been no accompanying changes in the legal framework for JFM, and in almost all states it has been implemented through administrative orders, which do not take the existing infrastructure into consideration. Forests continue to be looked upon as the property of the Government and the delegation of management responsibilities to local communities is merely an administrative arrangement, falling short of any real devolution of power to local communities. As an illustration. Prakash Kashwan has looked into JFM from various points of view and highlights some other problems. One of the most important findings is that JFM has been allowed only in degraded forest, but in order to bring more communities under the ambit of such programme, communities have had to actively pursue degradation of natural forest. Furthermore, JFM requires the demarcation of boundaries between villages, often for the first time, but poor implementation has resulted in many boundary conflicts.

Joint Forest Management (JFM) in the state of Jharkand;

- JFM in the state of Jharkand is based on the premise that people are to blame for the destruction of the forest. The objective, therefore, is gradually to diminish people's dependence on forests.
- The Resolution to introduce JFM has been made unilaterally by the Forest Department. There is no scope for consultation with the people, no effort at consensus building, and no mechanism for dissemination of information.
- The resources required for villagers' needs are to be extracted from the protected forest according to the micro/annual-working plan. There is no mention of the existing rights of the villagers to use the produce from these forests. This is a gross violation of customary rights.
- A 33 per cent share in net profits was promised. Now the promise is to share 90 per cent of net profits with the forest dwellers. The profits however, are from commercial productions, which are not based on local needs, thus indicating the focus on commercialization of forest. The JFM Resolution in Jharkhand is simply an extension of the undivided Bihar Forest Policy (Jharkhand used to be part of Bihar before it became a separate state), which adopted Joint Forest Management in 1990. Jharkhand Forest Department adopted the JFM Resolution on 27th September 2001
- The Forest Department claims to have formed 2034 Village Forest Management and Conservation Committees and Eco-Development Committees managing an area of 5,589 sq. km. in a remarkably short time span. However, local people emphatically rejected JFM. Apparently, the figure is manipulated and incorrect.

Eviction of communities as Forest Encroachers

Indigenous peoples feel that measures that were put in place in the name of protecting the forest have only deprived them of their resources. The JFM initiatives have failed from the point of view of indigenous peoples, and there has also been widespread eviction of communities from forests.

The forest department uses any means available to alienate the forest communities from the forest, reduce their dependence on the forest in the name of forest protection and conservation, and invite private partners and multinational companies to negotiate directly with village committees to access forests for carbon trading, Clean Development Mechanisms and trade in eco-services.

The direction of forest management is thus very far from anything that takes traditional forest related knowledge into account or that recognises the rights of indigenous peoples to maintain their traditional forest practices; rather, it aims to separate them from the forests, and sets up an externally imposed process of commodification and marketing of forest goods.

Indigenous Peoples' Rights and Forest Management

The Government of India is not a signatory to the ILO Convention 169 and as such does not recognize indigenous peoples. It does recognise "Scheduled Tribes", but not all indigenous groups are scheduled. Some of the indigenous groups are recognized by the Government of India as Scheduled Tribes and enjoy a degree of special status. Some of the areas where the Scheduled Tribes live have been demarcated as Scheduled areas and under the Indian Constitution Sixth Scheduled provisions, Autonomous District Councils have been set up for local administration of the Scheduled Tribes.

In theory there are some measures to protect the interests of Scheduled Tribes. The Constitution empowers the President of India and state governors to withhold any law considered detrimental to tribal/indigenous peoples' interests in Scheduled Areas. Items V and VI of the Indian Constitution give special privileges to the Scheduled Tribes and the Panchayats Act (Extension to the Scheduled Areas), 1996 (PESA) is also designed to offer some protection. However, in practice most of the laws restrict the rights of and control of forest communities. Specifically, the Indian Forest Act (IFA), Forest Conservation Act (FCA) and Wildlife Protection Act (WLPA) continue to be used to hound forest dwellers.

JFM and CFM are present in most Indian states, but are simply plantation programmes related to timber and other economic products, which have nothing in essence to do with forests. They should include Village Forest Committees directly representing the community living in or using the forest. These programmes are operational to varying degrees in different places with varying degrees of participation of indigenous peoples depending upon the local situation. The involvement of indigenous peoples is focused on providing them with a share in the economic benefits and *de facto* also acts to wean them away from their dependence on forests. In other words, integral to these programmes is a subtle (in some cases blatant) redefinition of indigenous peoples' relationship to the forests. The degree of incorporation of TFRK is varied, if there is any at all; the expertise is provided by the forest department, sometimes in partnership with NGOs.

Indigenous peoples' rights to the forests are progressively minimized. In some cases, indigenous peoples do receive some economic benefits (though in others non-indigenous peoples dominate the Village Protection Committees), and in the context of the takeover of resources necessary for basic survival and consequent deprivation, these economic benefits at times are seen as a steady source of survival income. This in certain cases has superficially reduced tensions and conflicts between the forest department and the indigenous peoples as a sense of co-existence is manufactured. However, this also means that the forest department is able to stifle protests and resistance. In many places, where the indigenous peoples' political consciousness is well developed, they keep away from or oppose these programmes leading to conflicts, as has happened in parts of Madhya Pradesh.

India's report to UNFF3 (2002) describes measures taken to make use of indigenous capacity and local knowledge in participatory forest management. Clearly, the picture painted by this report is not matched by the situation on the ground.

POTENTIAL BENEFITS IN FOREST MANAGEMENT

Forests are central to human landscapes, and civilizations have vanished because they failed to protect forests that in turn protected fragile soils and water sources. Present-day forest managers have a responsibility to spare California a similar fate. Forests are popularly believed to reduce flooding and increase dry-season base flows, as romanticized in the novel "The Man Who Planted Trees". The scientific evidence for relations between forests and water supply, however, has not been as compelling. Research has shown that forests have had a limited role in flood protection and variable effects on total water yields and base flows. In contrast, several studies have convincingly demonstrated that forests protect water quality by protecting the land surface from erosion and filtering pollutants.

Winter snowpacks presently store enormous quantities of water in the forested watersheds of California. Water stored as snow is released slowly during spring snowmelt, in constrast to rainfall that runs rapidly to streams. Snowpacks therefore supply water when it is most needed by humans and the environment, and reduce the need for additional downstream dams and reservoirs.

The predicted shift towards more rain and less snow is of critical importance for water management because the existing water-development infrastructure relies on regulation of streamflow through gradual release of water during snowmelt. As snow is replaced by rain at mid-elevations, winter flood peaks will become larger and more frequent, and reservoir storage is likely to be exceeded in wet months when demand is low. Summer baseflows will be correspondingly lower when demand is high. These climate-driven impacts will likely lead to proposals for many new dams and reservoirs on forest streams, with their resulting environmental impacts.

Climate changes directly affect forests through increased drought stress, which makes trees more vulnerable to insect attack. Wildfires are also likely to increase in frequency, size, and severity as climate warms. These stresses on forests will affect their capacity to naturally regulate streamflow and buffer water quality. Many streams that are now perennial are likely to become intermittent with the resulting loss of riparian zones, aquatic habitats, and other beneficial uses of water that depend on perennial flows. Recent concerns with increased demand for water, extended drought, economic and environmental costs of new water-supply infrastructure, effects of water transfers on endangered species, and effects of climate change on water supply have raised the importance of forest management for protection and improvement of water resources. New research and public perceptions may provide opportunities that were not available in the past to forest and water managers. Although the current scientific consensus supports the perception of forests primarily as protectors of water quality, the potential for improvements in the availability of water should not be overlooked. Forest management activities that alter streamflow regimen to benefit downstream water users may be more successful than attempts to increase total water yield. The following sections discuss forest management actions that have potential for improving water resources in California.

MEADOW GROUNDWATER STORAGE

Meadows within the forests of California are small alluvial landforms that are located along streams in the mountainous headwaters of the state. Owing to geologic factors that favour meadow formation, most of California's meadows are in the Sierra Nevada, which has more than 10,000 meadows comprising a total of roughly 300,000 acres. In contrast to the surrounding terrain, meadows have gentle slopes and support flood-tolerant herbaceous plants and shrubs rather than conifers. Under natural conditions, meadows lack incised stream channels, and high flows spread across the meadow surfaces and recharge meadow aquifers. Meadows with intact vegetative cover act as natural reservoirs, regulating streamflow regimen through storage and release of snowmelt and rainfall run-off that passes over and through fine-grained, sod-covered meadow deposits. These meadows attenuate flood peaks and prolong dry-season base flows. The importance of meadows in regulating streamflow is likely to increase as climate change results in a shift from snowmelt to rainfall-dominated run-off at mid-elevations in the Sierra Nevada.

Meadows also provide critical aquatic and riparian habitats for fish and wildlife. Montane meadows of the Sierra Nevada are very important to amphibians such as the mountain yellow-legged frog and birds including the willow flycatcher and Bell's vireo that utilize nesting habitats in meadows to successfully rear young. Kattelmann and Embry (1996) stated that about 20 per cent of the 400 Sierra Nevada terrestrial vertebrates are dependent on riparian areas. Field studies have shown that the health of benthic macroinvertebrate and fish communities is strongly correlated with riparian conditions. Most of the meadows throughout the Sierra Nevada were eroded by incised channels, or gullies, prior to 1940 as a result of early unrestricted livestock grazing, roadbuilding, railroad construction, elimination of beavers, and other causes. Although current activities on national forests are carefully managed to avoid damage to meadows, the effects of earlier practices remain on the landscape and will not heal without active restoration.

Eroded meadows lose their capacity to store and release water. Gullies convey and concentrate flood peaks more rapidly than well-vegetated meadow surfaces, and therefore aggravate downstream flooding and reduce recharge of meadow aquifers (Liang and others, 2006; Hammersmark and others, in review). Gullies drain groundwater stored in meadows, decreasing the amount of water available to sustain streams during dry summer months, although reduced evapotranspiration may offset drainage to an unknown extent. Channel erosion in meadows adds to stream sediment loads through bank erosion and headcut retreat, adversely affecting downstream water quality and reservoir capacity.

Erosion of meadows adversely affects aquatic and riparian habitats by altering meadow hydrology. Fish passage is not possible through incised channels that are dry for much of the summer months, and the reduced discharge of cold water stored in meadows is likely to increase downstream water temperatures, to the detriment of cold-water aquatic species such as trout.

Wildlife species that depend on wet-meadow plant communities are displaced when meadow water tables drop below rooting depths. According to the Sierra Nevada Ecosystem Project (1996), aquatic and riparian habitats are the most altered and impaired systems in the Sierra Nevada Province.

Drying of meadow soils allows invasion by drought-tolerant brush and conifer species that contribute to heavy fuels loading and add to the risk of catastrophic wildfires.

Loss of high-quality forage provided by wet-meadow sedges, rushes, and grasses decreases forage value for meadows that are grazed by livestock. Meadow restoration is a form of groundwater banking that can provide a wide array of ecological benefits in addition to enhancing water supplies. California's forests encompass the headwaters of the major rivers within the Sierra Nevada, and include thousands of meadows. A regional approach to meadow restoration could help to meet the State's needs for high-quality water and aquatic habitat and help offset the effects of climate change.

The USFS manages many Sierran meadows on national forest system lands, and has been actively working with partner agencies and organizations to restore the hydrologic, geomorphic, and biologic functions of meadows damaged by channel downcutting since 1940. Several projects using the new "plug and pond" approach have been successfully implemented in the past 10 years in the Shasta-Trinity, Plumas, Tahoe, and Sequioa National Forests, as well as the Lake Tahoe Basin Management Unit.

The "plug and pond" method involves preventing headcut migration, filling or plugging gullies, routing surface flows over the meadow surface, and raising meadow water tables. In most cases, livestock grazing has been temporarily halted during and after restoration projects. Permanent exclusion of livestock is not generally necessary to protect riparian resources if pastures are effectively managed to limit cattle numbers, distribution, and seasons of use. Like dams, meadow restoration does not create "new" water, but alters the temporal distribution of streamflow so that less water flows downstream during peak run-off periods in the winter and spring, when water is not in demand, and more more is released during the summer low-flow season, when demand is high.

Based on the limited available information and a reasonable range of assumptions, meadow restoration in the Sierra Nevada could increase the amount of groundwater stored in meadows by 50,000 to 500,000 ac-ft annually. The wide range in these estimates results from uncertainties in channel depths and specific yields of meadow alluvium. Increased groundwater storage in meadows would be likely to enhance summertime instream flows, a function that will become increasingly important owing to climate change.

Meadows are often considered to be "sponges" that absorb and release water. This analogy is appropriate for meadows that are supplied by streams and local snowmelt. Meadows that are supplied by groundwater flowing through surrounding hillslopes and underlying fractured bedrock, however, may more accurately be characterized as "valves" that regulate the discharge of regional groundwater as it flows through relatively low-permeability, organic-rich, fine-grained alluvium. Meadow restoration is likely to be most effective in prolonging the duration of base flows in meadows that act as "valves" in regional groundwater flow systems. On the basis of bedrock permeability, meadows that function as "valves" are more likely to be found in volcanic and weathered granitic watersheds than in glaciated granitic watersheds.

Meadow restoration offers a number of other benefits in addition to improving summertime streamflows. Some of these benefits are directly related to water resources, including attenuation of downstream flooding, improvements in downstream water quality, and reduction of reservoir siltation.

Other benefits are related to natural resources other than water, and include improvements in fish and wildlife habitat, protection of sensitive species, reduction of fuels loading, carbon sequestration, and improvements in range conditions. Alluvial valleys in mountainous areas throughout the world, including Africa, Australia, Europe, and South America, are faced with erosion and water-supply problems similar to those facing California's Sierra Nevada montane meadows.

Many of these alluvial valleys provide water, crops, and forage that sustain local communities and economies. Successful restoration of meadows in California could provide methodologies that are applicable to critical land and water degradation problems around the world.

Riparian Forests

Riparian forests are forested lands that are located immediately adjacent to streams, lakes, or other waterbodies. Riparian areas represent a transitional zone between aquatic and terrestrial habitats. These habitats are related to and influenced by surface or subsurface waters, especially the margins of streams, lakes, ponds, wetlands, and seeps. Boundaries between riparian and upland forests are not always distinct. The spatial extent of riparian areas varies laterally throughout the channel network and is strongly influenced by geomorphology. Riparian floodplains in forested watersheds may affect or be affected by hydrologic processes in several ways. Due to flow resistance associated with a forested stand condition, flood flow velocities are reduced considerably.

Additionally, when flow on the floodplain occurs, there is a large increase in the cross-sectional flow area. Once a stream rises above the floodplain surface, the slope of the line defining the relationship between river stage and discharge markedly decreases due to increased surface roughness and shallow water depth. In most instances, both the velocity and depth of water flowing outside the channel declines with distance away from the channel. This allows flood waters to recharge alluvial groundwater aquifers and attenuates downstream flood flows.

Floodplains are very important in providing habitat for riparian-dependent species. Floodplains are zones of very high biological diversity, having the highest biodiversity for both terrestrial and aquatic species of any part of the landscape at the watershed scale. In light of this high biodiversity, the California Forest Practice Rules require that native aquatic and riparian associated species and the beneficial functions of riparian zones must be maintained where they are in good condition, protected where they are threatened, and restored where they are impaired (in so far as is feasible). Harvest of coast redwoods on riparian floodplains in the northern Coast Range has been controversial because redwoods on alluvial flats have both high timber volume and unique ecological benefits.

Studies have shown that riparian forests can improve water quality. As surface run-off passes through riparian forests, materials carried in suspension are filtered by riparian trees that act as buffers to reduce the amount of sediment, nutrients, and pesticides entering water courses. Riparian trees also lower stream water temperatures through canopy shading. Shade is important for many fish species that can be adversely affected by elevated water temperatures at different life stages. In addition, riparian forests deliver wood to streams that contribute to pool formation and provide cover for fish.

Riparian forests are protected on federal, state, and private timberlands through riparian buffers that limit management actions such as timber harvesting and roadbuilding near streams. The width of riparian buffers, and restrictions on management activities within them, are based largely on land ownership. Within the national forests, riparian buffer widths vary based on provincial plan standards and guidelines. Riparian protection is most extensive for the 6 forests that operate under the Northwest Forest Plan. Fuels reduction within riparian buffers may be needed in some cases to reduce threats of catastrophic wildfires. Removal of trees from riparian buffers remains highly controversial, and forest management and regulatory agencies are carefully evaluating monitoring data particularly with regards to the use of mechanical equipment in streamside zones.

Some riparian forests are used for livestock grazing, usually within allotments that are primarily upland pasture. The availability of water and forage make riparian areas attractive to livestock, which can damage riparian forests through trampling, browsing, and contamination of streams with fecal material.

BMPs for range management and national forest standards and guidelines for riparian management are designed to protect riparian forests from damage by livestock. Although exclusion of cattle may be needed during and immediately after restoration of riparian forests, grazing strategies that minimize impacts on riparian forests through livestock numbers, distribution, and season of use can be used to eliminate the need for permanent fencing. Riparian forests include phreatophytes (trees dependent on availability of shallow groundwater, such as aspens, cottonwoods, and willows) as well as upland species such as oaks, maples, and conifers that utilize moisture above the water table. Riparian forests therefore may affect and be affected by channel incision and groundwater storage in much the same ways as meadows. Riparian trees can also be directly impacted by changes in channel geomorphology, including bed aggradation and channel widening that are frequently linked to upstream land uses, including forest harvesting and roadbuilding.

The water quantity and quality benefits provided by riparian forests can be preserved and enhanced through actions that maintain natural channel geomorphology. Protection of riparian forests therefore depends heavily on effective management of upland watersheds to prevent excessive run-off and sedimentation.

Vegetation Management

Management of forest vegetation to improve water supplies has a long history in the western United States. Early efforts attempted to reduce transpiration or increase snowpack by removal of trees. Most of these efforts had limited success. Previous studies have shown that increases in water yields resulting from vegetation management are highly variable and difficult to measure, and that treatments must remove at least 20 per cent of the vegetation to have a measurable effect on streamflow. Using computer simulation work, Troendle and others (2007) reported that an increase of 1 acre-foot of run-off theoretically could result for every 12 acres of forest thinning (fuels reduction). They suggested that the water yield response to large scale forest thinning in the northern Sierra Nevada forests would be short lived with a single treatment (perhaps only 15 years), but that an active management programme perpetuated over time could result in subtle increases in water yield. Studies have provided limited evidence that measurable water-yield increases have occurred in larger watersheds in the past. For example, Blanchard (1962—as cited by Zinke, 1987) calculated the cumulative effect of 30 years of logging on the South and Middle Forks of the Mokelumne River in the central Sierra Nevada. He reported that from 1930 to 1961, water yields from these watersheds gradually increased. Approximately 40,000 acres were logged during that time period.

Innovative approaches that utilize selective thinning of younger, smaller trees show some promise for limited improvement in streamflow regimen, as well as reducing fuel loading and increasing carbon sequestration. These treatments, however, also have potential to increase surface run-off and erosion from disturbed soils.

Fuels/Fire Management

Wildfires affect water resources through removal of vegetation and alteration of soils and ground cover. Removal of forest canopies by fire temporarily reduces transpiration and interception losses, and may increase streamflow until vegetative regrowth increases transpiration to or above pre-fire rates.

Burning of schlerophyllous vegetation can lead to development of hydrophobic soils that have reduced infiltration rates. Hydrophobic soils, in conjunction with the lack of ground cover remaining after fires, can increase surface run-off and erosion and reduce infiltration and baseflows. The effects of wildfires on infiltration and run-off are positively related to fire severity. Wildfires can also temporarily affect municipal water supplies when large quantities of water are needed to fight fires in wildland-urban interface areas. Although increased water yield is a potential impact of large, intense wildfires, it is generally not significant. Where 75 per cent to 100 per cent of the vegetative cover is removed, run-off may increase from 0.1 acre-foot per acre of burned watershed for basins receiving 15 inches of mean annual precipitation to 0.8 acre foot per acre burned for watersheds receiving 40 inches of mean annual precipitation). In forested areas, water-yield increases are minimal until basal area loss to fire exceeds 50 per cent.

The additional water yields that result from catastrophic wildfires are generally considered to have little value for water supply and hydroelectric energy generation. Almost all of the additional run-off occurs during the wet season and must be regulated for dry season use by surface reservoir storage. Typically, flows increase during large storm events when water is intentionally allowed to pass through reservoirs owing to flood-management concerns. Additionally, the additional water yield does not contribute to a dependable water supply, since the additional water is only a very temporary increase in yield. The occasional, short-term positive gains from increased water yield are more than offset by the frequent short and long-term negative impacts of increased peak flows, increased sedimentation and decreased water quality.

High fuel loads that develop in the absence of fire or fuel-reduction treatments eventually lead to catastrophic high-intensity stand-replacing fires that generate large volumes of eroded soil and ash, as well as nutrients such as nitrate nitrogen, ammonium nitrogen and phosphate phosphorus. Fire exclusion can also lead to negative effects on water quality owing to unnaturally large accumulations of forest litter that increase concentrations of nutrients in run-off from forests.

Fuels reduction projects, including thinning, mastication, and prescribed fires, are used to decrease the intensity, extent, and negative consequences of wildland fire. Forest management to reduce fire severity on national forests is currently administered under the National Fire Plan (NFP) and the Healthy Forest Initiative (HFI). About 70 per cent of the 20,000,000 acres of national forest system lands in California, or 14,000,000 acres, are in need of treatments to reduce fuel loads to natural levels. The USFS is currently treating about 200,000 acres per year, while on average 400,000 acres are lost annually to wildfires (Rob Griffith, USFS, personal commun., 2008). The most effective fuels reduction treatments for reducing the spread and intensity of wildfires have been combinations of mechanical treatments and prescribed burning.

Fuel reduction projects can have adverse effects on water quality, but these effects are generally minor and temporary, and are far exceeded by the adverse effects of catastrophic wildfires. The adverse impacts of wildfire are generally much greater per unit of affected area than the impacts of fuels reduction projects, and also affect much larger areas than those included in fuels reduction treatments.

Firefighting tactics are increasingly being modified to protect water quality and aquatic organisms. Rapid restoration of areas disturbed by fire suppression actions is routinely included in suppression efforts. Following fire containment, burned areas are assessed and, if necessary, treated to prevent adverse effects on water quality and other resources. The USFS uses its Burned Area Emergency Response (BAER) programme to restore areas burned by wildfires, while CAL FIRE participates in the State Emergency Assessment Team (SEAT) programme on state and private lands. These programmes focus primarily on erosion prevention, as well as the re-establishment of forest vegetation. The BAER and SEAT programmes will become increasingly important owing to projections of higher frequency and intensity of wildfires related to climate change.

Road Management

Watershed research and monitoring work in California have shown that unpaved roads and road stream crossings are usually the dominant source of management-related sedimentation in forested environments. Past work has shown that a relatively small proportion of the total road length produces most of the road-related sediment delivered to streams. Road-related sediment-transport processes include surface erosion, gullying, and mass wasting. Sediment may be delivered to streams either episodically when roads or road-stream crossings catastrophically fail, or chronically due to incremental surface erosion. Excessive sedimentation associated with roads is of concern because of potential impacts on stream habitat and water quality. Thousands of miles of roads have been constructed through forests in California, primarily to provide access for timber harvests.

The eighteen National Forests in California alone contain about 50,000 miles of forest roads. Of these, roughly 20,000 miles may no longer be needed for their original purposes. Private forestlands contain many additional thousands of miles. Forest roads intercept surface and subsurface flows, adding to flood peaks and decreasing recession flows. Stream crossings are vulnerable to damage by high flows and can divert streams from their natural courses, resulting in serious erosion and water-quality problems.

Many of the roads on private timberlands were built decades ago to very low design standards, often in environmentally sensitive locations such as unstable hillslopes and riparian areas. A significant number of the older roads are part of the current road network, while others were neglected and abandoned with no consideration of ongoing erosional impacts. These "legacy" roads are particularly susceptible to catastrophic failure during high magnitude, low frequency storm events, such as the one in 1997 that caused extensive flooding throughout a large part of northern and central California (Furniss and others, 1998, Madej 2001).

Similarly, much of the existing national forest road network was constructed under standards less protective of water resources than present standards, and maintenance of roads has not been completely effective at preventing run-off interception and erosion. Many roads constructed for timber harvest are no longer needed for their original purpose. Adverse hydrologic and water-quality impacts of roads can be reduced by upgrading and replacing culverts, outsloping road treads, and installation of waterbars and grade dips at stream crossings. Roads no longer necessary for resource management or recreation can be effectively decommissioned by removal of fills at stream crossings and partial or total outsloping of road treads, including cuts and fills.

Detailed field surveys are the main tool available to identify the road segments of greatest concern. Public and private landowners in California are actively inventorying their road networks, prioritizing road segments requiring road improvement or decommissioning work, and completing projects. A considerable amount of road upgrade work has been completed to date with both public and private financing. While there are short-term impacts associated with road decommissioning, particularly at road-stream crossings, road treatments will reduce the long-term sediment production from older roads. Improved operator practices are required for road-stream crossing installation at high-risk sites and at decommissioned crossing sites.

Urban Forestry

Trees planted in city parks, lots, private residences, and along streets collectively form urban forests. Although urban forests are not managed specifically for natural resource production or conservation, they have environmental benefits that extend well beyond aesthetics. Trees reduce stormwater run-off by intercepting rainwater on leaves, branches, and tree trunks. Rainwater intercepted by trees soaks into the ground, reducing peak flows, surface run-off, erosion, and contaminant transport.

Canopy interception from urban trees changes run-off quantity and pollutant loads in several ways:

1. Evapotranspiration – increases soil water storage potential;
2. Interception – reduces the volume and timing of run-off;
3. Infiltration – The root system of trees can increase soil infiltration rates;
4. Erosion – Through interception trees reduce the soil impacts from raindrops.

Trees in combination with other types of landscaping can be very effective in intercepting rainfall and filtering pollutants. Comprehensive urban forest management can lead to improved stormwater control and promote efficient water use in urban areas.

FOREST BASED INDUSTRIAL DEVELOPMENT

The objective of the present study is to analyse and understand the relative dependency of different forest-based industries on forests and their survival strategies in the context of diminishing forest cover and non-availability of timber in the Western Ghats of Karnataka, India. A stratified sampling technique was adopted to select small-scale industries and a census method was used to choose medium and large-scale industries. The study found that Government policy was mainly responsible for the fast growth of forest-based industries in the selected districts.

Forests provide raw materials to large, medium, and small-scale industries. Hence, forests are a source of life for the forest-based industries. As the New Forest Policy (1988 and 1992) restricts the felling of green trees, the industries are now facing the problem of raw material shortages. Among the industries taken up for the study the medium and large-scale industries have better prospects than small industries. Many small wood-based enterprises, cart manufacture for example, experience problems. More afforestation and a commitment by the forest department to supply forest produce to industries is the only way for sustained survival of small-scale wood-based industries in the Western Ghats of Karnataka, India.

Timber industry in India was initially confined to produce building materials, agriculture implements, bullock carts, and railway sleepers. Forest based industries were encouraged because of its rural identity and its ability to solve the problem of unemployment and poverty. In addition, the policy makers had also perceived that natural resource based industrial development is a pre-requisite for the economic development of less developed countries like India. Hence, the Government, both at the Centre and State encouraged establishment of small, medium, and large-scale forest based industries in the region. Due to this policy, the country witnessed heavy pressure of industries on forests for raw material purpose.

There is no industrial forestry activity apart from small scale charcoal production (largely for tourist and domestic recreation pursuits), extraction of wattle for fish and lobster traps, and some use of mahogany and white cedar for boat construction and fence posts. Estimates are that 1135 tonnes/year of charcoal and 4000 fence posts/year are produced, but there are no supporting data. The wood related manufacturing sector is also small, but again there are no records. Some opportunity may exist for handicraft material for carving, but technical training would be required to enhance growth in this area, particularly to utilize the limbs pruned from urban mahogany trees, some of which are dying or so badly mutilated that they should be felled and replaced. In addition, there is potential for non-wood products of agroforestry such as fruits and nuts, fodder and water.

Neem has been quite successful in Antigua and Barbuda and is widespread, mainly along road-sides or fence lines. It seems to have withstood the effects of hurricane Hugo in 1989, and attention should be given to determining the utilization prospects for this valuable fast-growing species. It is being planted elsewhere in the world and produces a wide range of products. These include the extract from the seed kernel and seed coat to yield neem oil, which is used as a lubricant, insect repellant and insecticide and in the manufacture of soaps and cosmetics.

The seedcake residue is a valuable organic manure and cattle feed, while the root exudate has medicinal properties; the bark yields an industrial gum and the leaves are used as a green manure. The timber is also useful for lumber and fuelwood. Antigua and Barbuda has an excellent opportunity to investigate the industrial potential of this tree which could attract investment to the country.

FUELWOOD AND ENERGY

In spite of the relatively high per capita income levels in Antigua and Barbuda, there is still some consumption of charcoal. Unfortunately, no reliable records exist to indicate the levels of production and consumption. Charcoal burning does not presently constitute a threat to forest resources as suitable material is already available from pastures in the form of Acacia. This tree is a pest which threatens the health of cattle, as the thorns cause severe skin lesions, exposing the animals to the risk of disease. The GOAB is obviously aware of the continuing world energy crisis, and a case could therefore be made for encouraging the use of alternate sources of energy to save foreign exchange and to utilize natural products generated in the course of development initiatives such as agroforestry and watershed rehabilitation.

Conservation of Forest Ecosystems

In spite of its small size, the island possesses considerable wildlife and ecosystem diversity, much of which could enhance the growth of tourism but, more importantly, could awaken society's consciousness and pride in its natural environment. The Country Environmental Profile has recommended specific elements to be included in a system of national parks and protected areas, and these should be expanded to include coastal zones and forested areas.

Negative development pressure on the coastal zone can compromise the future ability of the country to sustain long-term socio-economic growth. All development proposals should be carefully reassessed to avoid adverse environmental impact. In the meantime, interim action can be taken to stimulate financing of resource management through collection of fees by local guides and monitoring of visitor levels and impact. Resource inventories are proposed under the NFAP projects.

Institutions

The main institutions concerned with forestry and environmental management include the Forestry and Fisheries Divisions, the Development Control Authority (DCA), the National Parks Authority (NPA) and NGOs, primarily the Environmental Awareness Group (EAG). Under the present tight fiscal policies, the public agencies are restricted in their ability to execute programmes. There are no funds available to undertake work programmes in forestry or agroforestry, to maintain monitoring of fisheries and fishermen, nor to expand the capability of the NPA. The NGOs led by the EAG, are particularly knowledgeable and committed, and though lacking in resources, they are quite resourceful. They have played a significant role in raising environmental consciousness, although their confrontational approach has been met with official suspicion. However, they are now widely accepted in official circles that recognize the desirability of closer collaboration. This has led to the appointment of the Historical, Conservation and Environmental Commission (HCEC), aimed at improved dialogue and coordination within the public sector. However, the general institutional situation of the public sector is weak in implementing capacity and performance, and the NGOs are still skeptical that the Government will change its ways.

The NPA is backed by strong legislation which provides for inter-sectoral linkages to consult and cooperate with agencies having similar aims and objectives. It also has the power to delegate its functions and this could be done through the Barbuda Council, Forestry and Fisheries Divisions, the EAG and other NGOs.

The Barbuda Council was established by the Barbuda Local Government Act (No. 15, 1976), to manage Barbuda's government affairs. The Council consists of 9 members who elect a chairman and secretary from among them. Proposals for capital and special expenditure are required to be submitted to the Ministry of Home Affairs (Central Government) who forward them to the Ministry of Finance. The Council has powers to make bye-laws for development on the Island.

The conservation and development of forest resources can only be implemented by capable resource management institutions, properly staffed and financed, and provided with the necessary authority for efficient execution of their responsibilities. Unfortunately, the lead agencies in Antigua and Barbuda do not seem to be adequately equipped at present, and will require considerable strengthening to improve their management capacity.

With the recent return of the country's first professional forester, the Government now possesses the nucleus of trained forestry personnel. The Forestry Division can now be empowered with the following responsibilities and functions:

(a) protection of the forest estate, wildlife sanctuaries and protected areas;
(b) afforestation and forest management in watershed areas;
(c) cooperation with private and Government agencies on urban forestry programmes (including the Botanical Gardens), forestry education and information;
(d) assessment and conservation of biodiversity and promotion of ecotourism;
(c) generation of self sustaining activities utilizing forest and non-wood products;
(f) preparation of budget plans and annual performance reports.

The following steps are therefore recommended for official approval and action:

(i) adoption of a formal national forest policy and appointment of the Chief Forestry and Wildlife Officer;
(ii) declaration and gazetting of the forest estate, followed by survey and demarcation of forest reserve boundaries;

(iii) preparation and approval of a baseline work plan for the Division, including job descriptions for all staff;
(iv) transfer of game wardens, caretakers and necessary staff to the budget of the Division;
(v) assignment of responsibility for supervising the maintenance of the Victoria Park and Botanical Gardens;
(vi) assignment of office accommodation, vehicles and equipment to enable efficient performance of the Divisions functions.

After establishment of the Division, there are a number of basic activities which it can undertake, namely:

- planning, layout and implementation of a forest recreation centre at Wallings;
- planning and implementation of urban forestry based on the Botanical Gardens;
- field assessment of planted forested trees, enumeration, diameter and height measurements, site conditions;
- assessment of charcoal production from leucaena at Dunbars;
- continued cooperation with Fisheries Division and NGOs in mangrove and wildlife monitoring;
- initiation of contacts with NPA to discuss areas of collaboration;
- selection of possible reforestation sites and planning proposals for employment absorption.

These activities can be undertaken with little or no new financing and are within the technical capability of the trained staff. The results will soon be evident if tasks are approached in a business-like manner and with an eye to the future.

PLAN OF ACTION

Based on the results of the Country Mission, and on the working documents prepared by various consultants, a Plan of Action is proposed for the development of the country's forestry subsector. The National Forestry Action Plan (NFAP) proposes elements of national development policies, a long-term policy for the sub-sector, together with priority area policies and objectives on which specific action projects are proposed for a five-year period.

Mindful of the country's present financial situation, the Plan proposes redeployment of the existing personnel and budget as the Government's initial investment in the sub-sector to attract donor support in setting a sound foundation for sustainability. The plan includes the normal work of the Forestry Division which will also function as lead agency in the implementation of three projects. The NFAP is estimated to cost a total of US $3.0 million of which US $2.2 million would be contributed by international donors while the remaining US $0.8 million would be contributed by Antigua and Barbuda, mainly in kind.

National Development Policies

The adoption of national development policies for land use planning, land allocation and sustainable resources management are recommended for enhancing the contribution which forestry and trees can make to socio-economic development and environmental stability.

Long-term Subsector Policy and Objectives

59. It is proposed that the long-term Forest Policy of Antigua and Barbuda should be:

(a) to protect the country's natural forest resources;
(b) to promote the development of forests and tree cover on suitable lands;
(c) to encourage the application of agroforestry in planned land use for rural development;
(d) to pursue opportunities for utilization of forest and other products.

In pursuing this forest policy, the Government will seek to ensure:

- the wise use and proper management of the island's limited forest and natural resources;
- the preservation of representative samples of the island's natural forest and vegetation resources and the maintenance of the ecological balance;
- the provision of opportunities for the public for scientific, educational and recreational activities to enhance knowledge of forest ecosystems, research in potential utilization of the gene pool and tourism benefits;
- the stimulation of industries capable of boosting the island's economy, based on products derived from forests and trees.

Therefore, the proposed objective of the NFAP is to establish and manage a national forest estate on Government and private lands. This will ensure conservation of soil and water resources, protection of forest ecosystems and production of wood and related products for social and economic benefits on a sustainable basis.

Priority Area Policies and Projects

In the medium term, the NFAP will be guided by the priority area policies and strategies outlined below in the implementation of the respective action projects. These policies are all inter-related and represent the minimum actions required to effect the fundamental changes necessary towards achieving the basic aims of the new long-term forest policy.

Institutions-to provide effective environmental legislation and capable institutions for proper administration and sustained development of forestry and related programmes, with the participation of local people and interest groups.

The purpose of this policy is to improve the environmental management-capacity of the country and to install multi-sectoral and participatory mechanisms for planning and monitoring of forestry and related programmes. This aspect is of the highest priority if the national long-term policy and objective is to be attained. Accordingly, the following project is proposed for immediate attention:

Country Capacity for NFAP Implementation

The objective of the project is to build capacity in the lead agencies to ensure initiation and co-ordination of NFAP implementation, sustainable natural resources administration and management, project preparation and management, public education and extension, and environmental management for conservation of natural resources.

The project includes specific provision for the survey and demarcation of forest reserve boundaries in order to create a forest estate.

Forestry in Land Use-the proposed policy recognizes the multiple benefits which accrue from existing tree cover and seeks to promote forestry and agroforestry as appropriate land uses for scrub lands and areas in need of conservation.

In pursuing this policy, the aim will be to allocate an adequate area of suitable land to forestry and agroforestry to ensure conservation of soil and water resources, sustainable income and employment for rural people, and integrated crops and livestock production. This will require a programme of land use survey, classification and planning to effect proper subdivision and lease-hold settlements as incentives towards encouraging greater tree coverage on farms.

Forest and Watershed Rehabilitation

The objective of this project is to establish tree cover on degraded watersheds and sand-mined lands in order to reduce erosion, improve water retention and enhance rural development through agroforestry practices and reforestation.

Forest based industrial development-the proposed policy seeks to ensure production of future supplies of wood raw material and utilization of existing mature timber for industrial purposes, while maintaining the amenity value of trees in urban areas.

Utilization of Neem

The objective of this project is to study the prospects for utilizing various products of the neem tree with a view to industrial development.

Fuelwood and energy-the proposed policy seeks t-explore opportunities to utilize fuelwood for energy purposes and to promote feasible wood energy applications in appropriate industries. The aim would be to develop a sustained demand for wood energy by domestic and industrial users, whether as fuelwood or as charcoal, both to promote the use of agroforestry products and to conserve other energy sources.

The objective of this project is to determine the production and demand for charcoal and to evaluate the opportunity to expand the market.

Conservation of forest ecosystems-this policy seeks to maintain biological diversity and to protect forest ecosystems and natural habitats, while sustaining the flow of social, economic and environmental benefits derived from forests for nationals and tourists.

The aim is to increase public awareness of environmental problems and the need to conserve the natural resources of the island as part of the nation's heritage. In the absence of a coordinated system of management of existing habitats, the following project is extremely important.

SITUATIONS OF FOREST MANAGEMENT FOR CARBON COMPETE

The sites which were selected for this study are all in places where historically, degradation is the main process by which forest carbon was being lost. These are zones of rather low land value, where there is no obvious competition for alternative land use such as agriculture, because of the terrain, infrastructure such as irrigation, or because of the distance from markets. In such areas, the opportunity costs are low, and a small reward for carbon stock increase or for reduced carbon emissions may represent an attractive financial opportunity. In zones close to cities or in areas of high agricultural potential, there is more likelihood of wholesale land clearance (deforestation), and the value of carbon is probably not sufficient to counteract these processes.

The areas studied were all also managed by communities, rather than individual landowners or land users. This means that the unit forest size is in the range of 50 to 600 hectares, an area in which carbon stock can easily be measured and monitored by two or three people in two or three days. This offers considerable economies of scale over individual landholdings, where each individual would have to be trained, or for each of which a special arrangement would have to be made to carry out the stock assessments. It is also the case that forest degradation is related to uncontrolled community land uses (open access behaviour) while deforestation is more likely to be the result of individual land management decisions, as for example in the Amazon frontier where individual settlers (legal and illegal) move in and clear forest for pasture or for cropping, and the West African rain forest belt where timber companies may (legally or illegally) engage in clear felling. Thus there are several reasons why community forest management is particularly well placed as regards crediting for reduced emissions from degradation.

LOCAL TRANSACTION COSTS OF MEASUREMENT

The case studies have also demonstrated the utility of handheld computers with GIS/GPS equipment which make possible accurate mapping of the forest areas and which facilitate the storage of data on carbon stock. This seemingly 'high-tech' approach was found to be very suited to the local conditions, and village people with only a few years of primary education were able to use it after only a day's training (most of them were quite experienced in using mobile phones, which, anno 2006, are common even in the most remote villages). Indeed, local people were quick to recognize the power of such a mapping system and the additional uses to which it could be put (resolution of boundary problems with neighbouring villages etc.).

Of course, maintenance of the equipment is another matter, including recharging of the computer batteries (most of the villages were off the grid), scanning into the computer a suitable basemap, and setting out the sampling plots. For these activities it is clear that an NGO or private sector organization with some technical expertise is essential. Given this, and the cost of the computers (about $500, with a similar amount for the software) it is also clear that if the carbon stock change assessments are to be made in a cost effective manner, community forests would have to be clustered into groups, with an NGO or umbrella organization with one set of equipment assisting in perhaps 20 or 30 such forests.

Financing the Carbon

If community forest management is to be employed as a means for reducing emissions from deforestation, and particularly from degradation, in developing countries, then mechanisms to support this will be required, of which financial mechanisms will play a central role. There are issues as regards both international channels for finance and local channels for finance within the countries concerned. Here these are considered from the point of view of community forest management and how such communities could be rewarded for involvement in reducing emissions from deforestation.

International Finance Mechanisms

At the international level, two distinct modes for finance of 'reduced emissions from deforestation' are under discussion, although some combination or hybrid would be possible. The first draws inspiration from the Kyoto flexible mechanisms, and might be placed under the Kyoto Protocol through amendment or in a re-negotiated agreement relating to the second commitment period, which is expected to cover the years 2012-2017. In this approach market mechanisms are central; carbon credits are issued per ton of carbon emission reduced or sequestered, and in principle payment is made ex-post on the basis of this output. The idea is that these can be used directly to meet emission reduction goals. Within this general model of finance tied to carbon credits, there could be two possible ways in which this could be organized: either at a project level, as in the current CDM, with a project specific baseline representing the 'business as usual' rate of deforestation, or at a sectoral level, as in the proposal for 'Compensated Reductions', Party voluntarily accepts a national level target as regards emissions from deforestation, and the baseline is based on national rates of deforestation in the recent past. Either way, any reduction in the observed rate of deforestation compared to the baseline would be translated into tons of carbon, which would have to be verified and certified in some way before they can be sold. The important difference between these two versions is that in the first, the actors on the ground who are responsible for the reductions are directly involved in the deal, as with any CDM; in the second, the marketing deal is with the nation state, and nation state itself would decide how to distribute incentives or payments to encourage the actors on the ground to cooperate in reducing deforestation, or in what other ways the funds generated were to be used.

The second, quite different model for international finance is more in line with a traditional ODA approach to forestry. Financial assistance could be pledged to support efforts to counter deforestation, with a view to reducing emissions but without a direct or quantitative link to the number of tons of carbon saved, and without a direct link to the reduction targets. Such an agreement might fall directly under the UNFCCC rather than the Protocol, or indeed under another international agreement relating to forestry. Funds would be made available by states to support technical assistance and training, forest monitoring and inventory work, and other development activities with a view to helping developing countries implement policies and measures to effectively counter current rates of deforestation and degradation.

There are of course advantages and disadvantages relating to each of these models, which are much under debate at the moment. Many observers feel that ODA funds for forestry have had limited effectiveness as regards reducing deforestation in the past. Moreover they fear that if this model is used, the ODA payments will remain voluntary and not be forthcoming in large enough amounts to really change the current situation. It can be argued that only if payments are linked to legal and obligatory targets (as is the case with carbon reductions under the Kyoto Protocol) will there be sufficient pressure on countries to contribute the funds that would be required, rather than just making token payments.

There is also the understanding that a market system will be the most efficient in selecting the most economical carbon mitigation opportunities. On the other hand, the causes of deforestation and degradation are not simple; they result from combinations of many different factors, many of which cannot be tackled directly or individually. A holistic, developmental approach which provides opportunities for alternative livelihoods may be the best way to deal with the problem, but to relate this directly to observable reductions in emissions could be very difficult indeed, given the many drivers and causes at work, and the variations in this in different parts of the world and indeed in different parts of any country. It can be argued therefore that it makes little sense to fund reducing deforestation on the basis of simple carbon output.

Proponents of the market-based approaches take the view that the only means to stimulate real and sufficient investment by countries is to tie this to performance and to binding caps of some sort. The current reduction quotas (average of 5.2 per cent reductions over 1990 emissions) were negotiated before deforestation was considered as a CDM option, and clearly if reducing deforestation were to be admitted as a mitigation option in a Post-Kyoto regime, these caps would have to be re-negotiated, otherwise the market value of carbon would be threatened. Some have proposed that there should be a two target system, one for reductions in fossil fuel emissions, and a separate, but equally binding one, for bio-carbon emissions, including those from deforestation.

The advantage of carbon credits tied to projects is that the savings can be pinpointed easily to particular project activities and investments, as in any CDM arrangement. The major disadvantage and difficulty of including deforestation under the CDM approach is that it is very subject to leakage, through displacement of the deforestation activities to other sites, which is virtually impossible to avoid and very difficult to account for. For that reason a national approach in which the average rate of deforestation over a whole country is measured, rather than individual sites, is much to be preferred (as in, for example, the Compensated Reduction approach). This could, where necessary and sensible, be modified to refer to particular regions within a country. It could in principle also be expanded to cover a multi-nation region (to account for cross-border leakage which is common in many places) although this would make for a more complicated international agreement as regards sharing the credits.

Finance Mechanisms at the National Level

Whichever of the two basic models is eventually selected – a market based, carbon credit system or a system of greatly increased ODA financing to the forestry sector, focusing on reducing deforestation – there remains the question of how such funding is deployed within the country itself. It is noted that many countries have had considerable difficulty in controlling rates of deforestation in the past, not least because of increasing demand for timber products globally, but also because of internal pressures and competition for the use of land. In many cases the economic rent on retaining forest is so much lower than the potential rent from other activities that it is virtually impossible to prevent such shifts. This has of course a lot to do with the fact that the 'true' value of forest (its long term, environmental, intrinsic, and global value) is not reflected in the market system which drives such clearance. Policy mechanisms that can be used to control deforestation and degradation of forest within a country fall into three general categories, in common parlance referred to as 'sticks, carrots and sermons'.

'Sticks' are punitive measures designed to discourage activities leading to loss of forest; they include fines and other punishments for those who infringe laws and regulations designed to protect it. 'Carrots' are positive incentives such as payments for environmental services, or other rewards for not destroying forest. 'Sermons' refers to a wide range of informational activities which in different countries may be referred to as 'raising awareness' or 'education' of local people about the value of forest, or 'motivating' people to reduce their forest-destructive practices, through persuasion. Naturally, forest policy can rest on a combination of carrots, sticks and sermons.

Finance is required for measures in each of these categories, as well as to monitor closely the actual situation as regards deforestation/degradation. Clearly, whatever the package or mix of measures selected, some finance will be needed centrally to pay for the overall management and for activities that need input from the centre, while other finance will need to be distributed, particularly in the case of 'carrots', but also for the implementation of measures of the 'stick' and 'sermons' sort. The appropriate balance will be different in every country. Table gives a sketch of some of the possibilities. Here it is important to consider deforestation separately from degradation, since the two processes may have quite different drivers, and thus may require quite different counter-measures. Apart from other reasons, as already noted degradation often needs to be tackled through an organization at community level, since it affects the common property resources, while deforestation may be more often associated with individual or state land holdings and would need a different organizational approach.

THE FATE OF CARBON FUNDS IN THE FUTURE

If the local community were to be rewarded in financial terms for the carbon saved as a result of their forest management, would principles of equality hold, and would the poorer and less powerful part of the population, and women, benefit at all? The preliminary findings from the case study in Baghmara Buffer Zone Community Forest as regards the current distribution of benefits indicate that particularly as regards financial benefits, it is the richer parts of the population who gain most, even though most of the poorer people (Musahar excepted), and women, get a fair share of the products in terms of fodder and firewood. This outcome is not surprising since it is the men of higher caste and income that get to make the main decisions, despite the idea that the FUGs are supposed to be run on democratic lines. Whether this pattern would be repeated if a greater financial reward is entered into the system through sale of carbon sequestered or deforestation avoided, is not entirely clear. For example, one of the main reasons why the richer families benefit is because they are able to take loans for certain equipment from the community forestry fund; they have the means to match loans and collateral against the repayment. If money for carbon were not handled in the form of loans but (at least in part) distributed to members directly as an annual payment, then this problem should be overcome, and indeed the poor people would stand to earn a welcome, if small, additional income. It remains to be seen whether rules on membership would be tightened to limit membership in some way, if the financial rewards from carbon credits were considerable. At present membership is all inclusive. All this implies is that if equity goals are to be taken seriously, some serious consideration needs to be made regarding how the whole system of rules and procedures for internal payment of carbon services is to be designed, and that particular attention needs to be paid to how the needs and rights of the "weaker groups" will be guaranteed.

3

Population Genetics as a Basis for Selection

GENETIC STRUCTURE OF TREE POPULATIONS AS DETERMINING GENETIC PARAMETERS

FACTORS OF EVOLUTION AND THEIR EFFECTS ON POPULATIONS OF FOREST TREES

There has been remarkable progress in population genetics and the genetics of quantitative characters during the last few decades, and the results and methods have been widely used in forest genetics. This chapter explains how these results and methods can be applied to the selective breeding of forest trees.

The problem dealt with can be stated in the following way. Consider an economically important tree species with a given range; it is known that there are genetic differences between the subpopulation means of that species, and among trees from the same subpopulation; the intention is to use selective breeding to exploit for practical forestry:

1. The genetic variability between the subpopulations;
2. The genetic variability among individual trees in the same subpopulation.

The first of these two problems will more fully explained in the third section of this chapter but, in discussing the second problem, variation between subpopulations must also be included because the trees the breeder works with are usually selected from different subpopulations.

Genetic differentiation between subpopulations ensues from evolutionary events which lead to the development of the species and continue to influence it.

The interest here is in four groups of evolutionary factors:

(*a*) Accidents of sampling;
(*b*) Natural selection;
(*c*) Migration;
(*d*) Introgression.

Introgression will not be considered in detail although it must not be neglected in some cases. There are other factors of evolution which will not be dealt with here.

Natural selection and accidents of sampling result in genetic diversification between subpopulations, but migration counteracts this. AU three factors are active at the same time, however, so that the forest geneticist is faced with a population structure which is one stage in a process set in space and time. Normally there is a condition of balance between all three factors, and this balance cannot be far from an equilibrium state which permits genetic differences between subpopulation means because of natural selection and accidents of sampling.

The forest geneticist, therefore, will rarely, if ever, have to deal with a random-mating population. On the contrary, the distribution of genes will show considerable deviations from the Hardy-Weinberg equilibrium. This holds true also for obligate cross-breeding species.

GENERAL MODEL OF POPULATION STRUCTURE FOR MOST TREE SPECKS

The variation between subpopulations covering the whole species range, or a large part of it, can be divided into two portions:

1. A component for variation between geographic regions;
2. Another component resulting from genetic differences between subpopulations within the same region.

The first component will to a great extent reflect the effects of natural selection, while the second may be caused by natural selection or accidents of sampling, or both factors.

In most cases the "between regions" component can also be separated by means of regression techniques (clinal variation, as demonstrated by Langlet, 1936 and in later papers). The remaining variance can then be interpreted in the manner proposed for the "within-regions" component.

It is assumed here that accidents of sampling are the main source of within-regions variance. If natural selection is also a source of within-regions variance, we must assume a random distribution of environments to deal with this variation in a similar manner as the accidents of sampling.

By adopting these suppositions it is permissible to assume that: *the subpopulations within a given region car be thought of as a random group of inbred lines derived from the same random mating or tease population.*

Tree breeders will always work for certain regions (such as the *Wuchsgebiet* of Germany and *plantagezonen* of Sweden). This model for a natural population of a tree species within the boundaries of a region gives the breeder a chance to work with a defined population, which is necessary if he wants to use the methods of quantitative genetics. The boundaries of the "region" in the sense of the definition given above are chosen by considering climatic conditions and the results of biosystematic investigations (Callaham, 1963). If necessary it is possible to correct the boundaries from the results of provenance tests.

Following the definition, the population within the "region" does not have a random mating structure. It is subdivided in several subpopulations differing more or less from the general mean. By assuming that the subpopulations are randomly derived from a common base population and are equally inbred, we may be in a position to overcome this difficulty, as will be shown later in this chapter.

EXCEPTIONAL POPULATION STRUCTURES IN FOREST, TREES

The population model of the foregoing paragraph does not apply if:

1. The species under question propagate" mostly vegetatively, as is the case with willows and poplars;
2. The species is self-fertilizing, as most probably happens with the species in tropical rain forests (Baker, 1959);
3. the natural structure of populations has been widely destroyed by artificial regeneration.

In all three cases it is necessary to determine by experiment the conditions with which one has to deal.

QUANTITATIVE GENETIC PARAMETERS IN FOREST TREE BREEDING

CONCEPT OF HERITABILITY

Consider now a quantitative character showing continuous variation. The expression of this character in a certain individual or in a family is determined by the "genotypic value" of the individual or the family mean, and by the "environmental deviation" by which the performance of the individual or the family differs from the average performance of the whole population, due to differences in environment. The phenotypic performance can be considered as a linear combination of genotypic value and environmental deviation:

$$P = G + E$$

If G and E are not correlated, the variance of the phenotypic values is also a linear combination of their variances:

$$V_P = V_G + V_E$$

The ratio,

$$h^2 = \frac{V_G}{V_P} = \frac{V_G}{V_G + V_E}$$

is called the "heritability" of the character in that particular population. Heritabilities can be interpreted as the regressions of genotypic on phenotypic values and this is the reason for the wide use of heritabilities in planning breeding programmes.

There are several kinds of heritability. If (V_G) stands for the total genetic variance in a population, calculation of the ratio gives the heritability "in the broad sense." If total genetic variance ($V_{G'}$) is replaced by the variance resulting from additive gene effects (V_A), the calculation of the ratio gives the heritability "in the narrow sense" (Lush, 1937 and later papers). In both cases the denominator is total phenotypic variance, which is the sum of total genetic variance (resulting from additive variance, dominance and epistasis of genes) and environmental variance.

Through mass selection and related breeding procedures only the additive genetic variance can be utilized and here narrow sense heritability must be used. Broad sense heritability can be employed if vegetative propagation of selected clones for immediate practical use is possible.

Heritability may be used for the solution of different problems:

1. The heritability that represents the genetic fraction of phenotypic variation between individuals within a population;
2. The heritability that represents the genetic fraction of phenotypic variation between families; provenances etc., in a particular field experiment.

In the first case, h^2 is a parameter of a population; in the second case, it is a parameter of a particular experiment.

Hanson (1961) defines as heritabilities the mathematical expressions which measure the expected progress of selection. His main argument is that the breeder considers heritability for that reason only. Mergen (1960) uses heritability as a measure for "rigidity of genetic control" of quantitative characters. This is another application noting that heritability is always related to a specific population or experiment: h^2 then expresses the "relative importance" of the genetic sources of variation. Finally, h^2 can be used as a measure for the efficiency of an experiment (R. L. Anderson, 1960). A list of heritabilities published up to 1963 is given by Hattemer (1963) who also draws attention to several problems in connection with heritability estimation in forest trees.

COMBINING ABILITY

The concept of combining ability has been defined by Sprague and Tatum (1942) for inbred lines of maize but it is possible to apply it to other material. Combining ability is not an attribute of an individual or a line, but an attribute of the individual or the line and the population of other individuals or lines being investigated. Sprague and Tatum defined combining ability using the model of a two-way analysis of variance which can be adapted to a set of diallel crosses. The yield Y_{ij}, of the progeny after crossing the i-th with the j-th individual or line is composed of the average effects (a_i and a_j) of both lines over all their combinations with the other individuals or lines of the experiment, and a specific deviation, denoted by s_{ij}; m being the total mean:

$$Y_{ij} = m + a_i + a_i + a_j + s_{ij}$$

The variance resulting from the $a_{i,s}$ and $a_{j,s}$ is then the variance of "general combining ability"; the variance of s_{ij} being that of "specific combining ability.".

If the parents are a sample drawn from a randommating population, the variances of general and specific combining ability are estimates of the additive genetic variance and the dominance variance respectively of that population (neglecting epistasis). When selecting several clones for a seed orchard the breeder is mainly interested in general combining ability because he can exploit only additive genetic variance by means of such seed orchard. In other oases, however, it might be possible to use specific combining ability in species such as *Betula* and *Alnus* which allow seed production on a large scale from only two clones.

GENETIC CORRELATIONS

Genetic correlation between two or more characters involves two consequences for the tree breeder: he can make use of this correlation by means of the so-called *Frühtest* or "early test" if the young trees show one of the correlated characters and the other one can only be scored on mature trees. On the other hand, genetic correlation may lead to changes in characters not selected for.

Correlation of phenotypic values does not prove the presence of genetic correlations. Because:

$$P = G + E$$

such a correlation may result from a correlation of environmental deviations. Hence, genetic correlation can only be estimated in experiments where G and E are given separately.

Genetic correlations can result from additive, dominant, and epistatic gene effects. Here concerned is only with correlations of "breeding values" which originate additive effects. This type results from pleiotropic gene action in which identical genes affect both characters. (There may also be other genes which affect only one of the two characters.)

Genetic correlation found in a first generation experiment with forest trees can be explained often in other ways. It might result from linkage or from selecting the material from a subdivided population. It is suspected that at least some of the reported cases of genetic correlations in forest trees belong to those two categories. It is difficult to determine the type of genetic correlation on the basis of a one-generation experiment.

In an experiment with two Japanese birch species, *Betula japonica* and *B. maximowicziana*, genetic correlation due to subdivision of populations could be excluded. This experiment consisted of a number of single-tree progenies from each of 12 and 16 subpopulations respectively, sampled from the overlapping species ranges. Two characters, time of flushing and cessation of annual growth, were assessed by scoring in a randomized block experiment with four replicates. Highly significant correlations were found within subpopulations and of subpopulation means in *B. maximowicziana,* but none in *B. japonica*. Also in *B. maximowicziana* a significant mean square of subpopulation means is left after elimination of the covariance. This gives the impression of a linkage correlation. That impression is strengthened by the fact that there are parallel clines for both characters in *B. maximowicziana*, but in *B. japonica* a cline exists only for growth cessation (Stern, 1903). Reference should also be made to Kimura (1960) for genetic systems leading to linkage.

Other genetic correlations in forest trees have been reported recently by H. van Buijtenen (1963) and Sakai and Hatakeyama (1963).

COEFFICIENT OF INBREEDING

A change of the mating system of a randomly mating population leads to inbreeding, which can be measured in a simple way by the "inbreeding coefficient" F of S. Wright (1921) and Malécot (1948). Inbreeding plays a role also in natural populations of forest trees. This has been frequently investigated since J. W. Wright's (1952) first statement (Langner, 1962, Wang et al., 1060, and others). In a population having the structure described earlier in this chapter, the variation between subpopulations offers a means of estimating the degree of inbreeding. According to S. Wright (1921), in such populations

$$2\ FV_A$$

is the expectation of genetic variance between subpopulations (V_A being the genetic variance due to additive gene effects) and

$$(1 - F)V_A$$

is the expectation of genetic variance within subpopulations, assuming that V_A is the preponderant part of V_G.

Very little is known about the numerical values F can attain. The data in Table below are taken from the two nursery experiments with the Japanese birch species mentioned above. *Betula japonica* is an "opportunistic" species according to MacArthur (1960) and is expected, therefore, to show a higher degree of inbreeding than the "equilibrium" species, *B. maximowicziana*.

Table. Inbreeding Coefficients In Species of Betula

Characteristics	Year	Inbreeding coefficients	F as estimated by the two equations given above
		B. japonica	B. maximowicziana
Cessation of growth	1960	0.1215	0.0506
	1961	0.1248	0.0100

Leafing out	1961	0.0844	0.0150
	1982	0.1202	0.0268
Height growth until	1961/1962	0.1542	0.0497
	1963	0.1424	0.0346
Leaf rust,			
scored in summer	1960	0.1024	–
scored in autumn	1960	0.1030	–
scored in autumn	1961	0.1745	–

It is possible to avoid the bias from inbreeding in the estimates of h^2 if the magnitude of F is approximately known. This bias varies with the procedure used to determine V_A and this should be taken into account when choosing the mating design, if the value of F for the population concerned is unknown.

The degree of subdivision of populations within a region is not at present known for any tree species. Only lately tests have been prepared or laid out which will give some information on it and those that are recorded are listed here.

All these experiments recognize that the forest tree breeder needs a more complete picture of the populations to which he wishes to apply the methods of quantitative genetics. The different experimental designs cannot be discussed here although they would reveal the different viewpoints from which such experiments can be started.

Because of the phenomenon known as "inbreeding depression," inbreeding is also important in a breeding programme. Only a limited number of individuals can be admitted in each generation and the individuals of later generations thereby become more and more closely related. There are two ways in which the breeder can counteract this development. He can either keep the population fairly large or he can control the pedigrees of the parents in each generation. A large amount of literature exists on inbreeding in forest trees, beginning with Sylvén's (1910) paper, and there is little doubt about its importance. Minimum numbers of clones have been calculated by Stern (1959) while Langner (1961) showed that inbreeding depression in *Larix* is a linear function of F, as was theoretically expected.

ESTIMATION OF GENETIC PARAMETERS

MATING DESIGNS

The quantitative genetic parameters of a population can be estimated only from field experiments. Unfortunately it is rarely possible to establish particular experiments for this purpose and it is therefore advisable, at the present time, to lay out field tests to serve several purposes. These are:

1. To obtain information on the general and specific combining ability of parent trees for use in seed orchards;

2. To get reliable estimates of genetic parameters;
3. To establish material for further breeding work.

In addition, the test areas must be sufficiently small to keep the costs within tolerable limits.

The first step in planning progeny tests is the choice of a suitable mating design. Mostly the design chosen will be a variant of the diallel table (Andersson, et al., 1961; Zobel and McElwee, 1960 and others). Indeed, a field test with progenies from a diallel cross gives all the necessary information and remains small enough if it is well planned.

The diallel can be complete or incomplete (Hinkelmann 1960). It can take the form of a test-series, containing a few tester clones and a high number of clones to be tested, say b testers B and c clones to be tested C. In such an experiment, planted in randomized blocks, the analysis of variance of plot means will be as shown in Table below.

Table. Form of Analysis of Variance of Plot Means

Source of variation	Degrees of freedom	Expectation mean square
Blocks	r - 1	$V_E + bcV_R$
General combining ability B	b - 1	$V_E + rV_{BC} + rcV_B$
General combining ability C	c - 1	$V_E + rV_{BC} + rbV_c$
Specific combining ability	(b - 1) (c - 1)	$V_E + rV_{BC}$
Error	(r - 1) (bc - 1)	V_E
Total	rbc - 1	

Assuming that the parents in both sets of clones are taken at random from a random-mating population, V_C provides an estimate of ¼ of the additive genetic variance V_A, and V_{BC} an estimate of ¼ V_D of that random-mating population (V_D being that part of the genetic variance due to the dominance effects of genes). If the origin of the material cannot be defined in a correct manner, V_C may be called the variance of general combining ability of clones of set C.

Accepting the population model given earlier in this chapter, and the selection of clones over a whole region, it can be said that at,

$$V_B = \frac{1+F}{4} V_A, \text{ and } V_{BC} = \frac{(1+F)^2}{4} V_D$$

Other situations will not be considered here.

Mating designs other than the diallel gross have been suggested. Gustafsson (1949) recommends a pollen mix as the tester parent. Assuming a random-mating population, the variance of family means then gives an estimate of ¼ V_A. Schröck in a personal communication uses a single tester as pollen source or female parent. The variance of family means is then ¼ V_A + ¼ V_D, provided that the population has a random mating structure.

Single tree progenies derived from open pollination should give the same between-families variance as Gustafsson's polycross. But here the assumption of random pollination seems to be rather precarious (Squillace *et al.*, 1961).

The tree breeder should choose mating designs carefully. Already at this early stage he is deciding on the kind of information he will get in the future, on the size of the field test, and on the possibilities of further work with progenies from the experiment.

Another method of obtaining estimates of V_A is given by the fact that the regression of the expression of a character in parents and offspring is completely due to V_A. Finally, Shrikhande (1957), Sakai and Hatakeyama (1963) and Freeman (1963) all present a new method enabling the tree breeder, under certain circumstances, to estimate genetic variances and genetic correlations in even-aged stands without raising progeny.

FIELD EXPERIMENTS

The choice of field design, plot size, and number of replicates all determine the efficiency of an experiment, that is, the amount of information that will be obtained. Thus far there are not many data on the efficiency of field tests with forest trees but the studies of J. W. Wright and Freeland (1960), Evans *et al.* (1961), and Strand (1955) all confirmed that the basic assumptions of Andersson et al. (1951) were correct.

For different characters different error variances are to be expected from the same test. The heritabilities given in Table below were obtained from a test with 100 provenances of *Picea abies*. The design was a simple lattice with 4 replications; plot size 5 by 5 plants; spacing 1.1 x 1.1 meters (3 feet 6 inches x 3 feet 6 inches).

Table. Heritabilities Derived From A Provenance - Test of Picea Abies

Character	Number of replications							
	1	2	3	4	5	6	7	8
Leafing out in 1962	87	93	95	96	97	98	98	98
Damage by spring frost 1961	93	97	98	98	99	99	98	99
Height growth 1962	39	67	66	72	77	79	82	84
Prolepsis formation 1962	39	56	66	72	70	79	82	84

One single replication would give 87 and 93 per cent of available information in the case of leafing out and frost damage, respectively; but height growth and prolepsis formation show much lower heritabilities. Only with 8 replications would 84 per cent heritability be achieved. These figures were obtained without consideration of the incomplete blocks. Eliminating the variance among them, the efficiency would reach 156 per cent against that of randomized blocks.

It is desirable that, following the proposals of Evans *et al.* (1961), all available results of field tests should be collected and assessed for information on the optimum size of plots, number of replications and so on.

Furthermore, there should be more uniformity trials similar to those of J. W. Wright and Freeland (1960) or Strand (1966) at different locations and with other tree species.

It should then be possible in a few years to accumulate sufficient material for planning field tests.

INTERACTIONS OF GENOTYPE AND ENVIRONMENT

The tree breeder works in a region or selects trees in regions which are to some degree heterogenous in soil and climate. Furthermore, the weather changes from year to year, and in years with extreme conditions the results of many previous years can be reduced in value. The breeder has to take into account the fact that different genotypes may react differently in different environments.

He will have to repeat his tests on several sites and, if possible, over several years. Such a test series, an A x B diallel over 1 location with r replications, each in a randomized block design, has the expectations shown in Table below.

Table. Form of Analysis of Interactions of Genotype and Environment

Source of variation	Expectation mean square
A	$V_E + rV_{ABL} + rlV_{AB} + rbV_{AL} + rblV_A$
B	$V_E + rV_{ABL} + rlV_{AB} + raV_{BL} + ralV_B$
A x B	$V_E + rV_{ABL} + rlV_{AB}$
L	$V_E + rV_{ABL} + rbV_{AL} + raV_{BL} + rabV_L$
A x L	$V_E + rV_{ABL} + rlV_{AB} + rbV_{AL} + rbV_A$
B x L	$V_E + rV_{ABL} + raV_{BL}$
A x B x L	$V_E + rV_{ABL}$
R within L	$V_E + abV_R$
E	V_E

The heritability must then be taken over all locations. For example, if the sample A consists of the clones to be tested:

$$h^2 = \frac{V_A}{V_A + V_{AL}/1 + V_{AB}/b + V_{ABL}/b1 + V_E/rb1}$$

The test sites should not be chosen at random as frequently happens. The most important site types should be preferred.

Unfortunately, there are few data about genotype x environment interactions in field experiments with material derived from selective breeding in forest trees, and these data are not very reliable.

COMPETITION

Competition is the mutual influence of trees within stands and it has the following consequences for tree breeders.

1. The selection of plus trees becomes unreliable for characters affected by competition.
2. Estimates of combining ability and genetic variances are biased to some degree because of competition within and between plots.

The phenomenon of competition has not been sufficiently understood until now, in spite of some fundamental papers referring to it (Sakai, 1955; Symp. Soc. Exp. Biol. Cambr. 1961).

Partly because of competition the heritabilities of growth obtained from parent-offspring regressions are considerably smaller than those from half-sib or full-sib analysis. It can be assumed that the latter heritabilities are more reliable.

Competition in progeny tests is difficult to assess. If the families are planted in plots it is mostly half-sibs or full-sibs that compete with each other. On the other hand, the progeny of a seed orchard for which selection has been made consists of practically unrelated individuals and the competition is of a different kind. For this reason most tree breeders dispense with the isolation strips and large plots usual in other branches of forest research.

The results of Snaydon (1961) show that it is necessary to consider the competitive ability of different genotypes. He found no differences between ecotypes from different soils if grown in plots. But an ecotype on its own soil was always superior when grown in competition with the others. Older experiments by Sukatschew (1927) point in the same direction. Sub division of populations within a region, therefore, may be caused partly by competition.

Recently, Toda (1963) has presented a new and promising method for the estimation of heritabilities in the presence of competition. There should be more basic research on this problem, which is of vital importance not only for forest tree breeding but also for other aspects of forest research and practice.

HOW TO ACHIEVE GREATEST GENETIC GAIN

MEANING OF GENETIC GAIN

A new population comes into existence by selection. The average genetic value of this population deviates from that of the original population. This difference is the "genetic gain" specific for the method and intensity of selection used in a particular breeding programme. It is, of course, the intention of every breeder to make this difference as large as possible.

Three factors determine the gain obtained or expected:

1. The genetic variance which is the basis of any selection;
2. The accuracy of estimates of breeding values either in selection of

superior individuals in a stand or of families in an experiment, as expressed by the heritability;

3. The intensity of selection, that is the difference between the population mean and the average of the selected individuals.

It can be shown that under certain conditions the genetic gain R is given by the following equation:

$$R = h^2 S$$

where S is the difference between the means of the original and selected populations. This is the basic formula for all estimates of expected genetic gains.

SELECTION INDEXES

The economic value of a cultivar is determined by several characters. Hence, it is necessary to look for a selection procedure that takes account of the complexity of the situation and gives the best possible results, not for a single character but for the whole complex of characters making up economic value. The most promising procedure seems to be the selection index and the most complete survey of the problem has been given by Le Roy (1960).

In constructing selection indexes it is necessary to combine the economic values of the separate characters, their heritabilities and their correlations with other characters of economical importance. It is understood that genetic correlations are only available after comprehensive progeny tests. Assume a selection programme in which the breeder selects for characters $x_1 ..., x_n$. It is further assumed that the heritabilities, the genetic correlations, and the economic weights of all characters are known. It is possible to combine all this information in a single "index" which takes the form of a multiple regression but is calculated in a different manner (see Le Roy 1960, for details):

$$I = b_1x_1 + b_2x_2 + ... b_nx_n$$

The $b_1 ... b_n$ are the weights of the single characters and a character of high economic importance has a high weight. But this can be counterbalanced by a low heritability or by negative genetic correlation with other characters of economic importance. The $b_i ... b_n$, therefore, are calculated using both economic and genetic information.

The equation for determining genetic gain applies also for index selection, that is selection for high values of I. Nevertheless the problem of determining the optimal selection procedure for a given set of economic weights and population parameters is difficult to solve in some instances. There have been successful attempts in animal breeding to approach it by using high speed computers and model populations. A similar approach might also be promising in solving certain problems of forest tree breeding.

The economic value of a character is certainly problematic; it alters with changing market conditions. But it should be possible to give an economic weighting to the several characters combined by the index. Fast growth, straight stems, good natural pruning and so on, are characters which are expected to maintain their values for a long time.

The demand to include as many characters as possible in the index is countermanded by the fact that the intensity of selection for the others is lowered by every new character considered. Therefore the breeder has to restrict the number of characters he is selecting for and this stresses the necessity of a deliberate weighting to choose the most important characters.

Thus far only one "real" selection index has been constructed for use in forest tree breeding by J. P. van Buijtenen and van Horn (1960). But the scoring systems used by most tree breeders are also selection indexes.

The construction of the indexes, and their continuous adjustment to market conditions, is one of the most important problems of tree breeders. A solution of that problem can only be obtained by close co-operation with forest economists.

Since the heritability of a character can be low in stands but high in a well-planned field experiment, the weight of this character will be different in a selection index calculated to maximize gain by plus tree selection when compared with a second index referring to the same character and to the same population but to selection on the basis of progeny tests. A good example of this is given by Nixon and Moffet (1963).

FACTORS DETERMINING GENETIC GAIN

It is the aim of every breeder to make the widest possible use of available facilities. Usually this is identical with the aim of achieving the largest generic gain (in the index values) per unit of time. With annual plants and also with many animals the unit of time is taken as one generation, but with forest trees the generation is a variable period. Its length is determined by:

(*a*) The time from pollination to seed collection, which may be two or even more years;

(*b*) The time from collecting seed to the end of the field tests, which will vary considerably;

(*c*) The time needed to apply the results of selection in forest practice.

The breeder tries, of course, to shorten each period as much as possible. He might be able to combine several measures during the different periods. Assuming that each step in selective breeding is followed by others, it becomes necessary to estimate also the possible progress of further steps. Therefore a multitude of facts besides the selection index must be taken into account since they also influence the success of a selection programme. It seems that we are at the beginning as far as this kind of consideration is concerned.

EARLY TESTING

Characters with high heritability in the narrow sense can be selected for by selection of plus trees. In such a case high selection differential guarantees fast progress through mass selection without progeny testing.

The situation is different when selection for several characters is reliable only when based on progeny tests. Because of the high costs of such tests, it is desirable to restrict them to practicable sizes, although the chance for high selection differential is probably lost in smaller tests. Attempts have therefore been made to use genetic correlations between characters of young plants and those scored in the later stages of tree development, for indirect selection.

The success of such indirect selection as compared to that of direct selection can be estimated as follows (Falconer, 1060):

$$\frac{CR_x}{R_x} = \frac{i_Y h_Y r_A S_{AX}}{i_X h_X S_{AX}}$$

In this equation CR_X stands for the correlated response to selection for y in x, R_X for the response to direct selection for x, i_K and i_X for selection intensities for y and x respectively, h_X and h_Y for the square roots of the heritabilities, r_A for the correlation of breeding values in x and y, and S_{AX} for the genetic standard deviation of breeding values of x. The statements about genetic correlations made earlier in this chapter also hold for "early tests." In particular, it must be remembered that correlations arising from linkage change from generation to generation, and that correlations arising from the subdivision of populations are of no use for indirect selection within subpopulations.

SEED ORCHARDS

During the last few years there have been discussions about the most appropriate type of seed orchard. In particular, J. W. Wright (1959) and Goddard and Brown (1961) have drawn attention to seedling seed orchards, established by discarding all but the best trees in the best progenies of progeny tests. This concept of a seed orchard is quite different from that of the clonal seed orchards developed particularly in Sweden.

The object of this chapter has been to show that a large amount of information is required for the deduction of the most profitable selection procedure. It would be surprising if sufficient data were available now for any selection programme in a forest tree species, and in particular:

(*a*) Heritabilities for stands and certain types of progeny tests;

(*b*) Genetic correlations of the most important characters;

(*c*) Effects of inbreeding;

(*d*) Genotype x environment interactions;

(*e*) Economic weights of important characters;

(*f*) Time needed for each step of the programme.

At present it is necessary to guess or estimate these data subjectively and this seems a doubtful procedure. An experimental procedure which leads to the selection of superior material and provides estimates of genetic parameters seems to be the most promising.

In addition it is hardly possible to assume a standard type of clonal seed orchard for all species and breeding programmes and the Swedish concept is a very flexible one. Much can be achieved by choosing a suitable timetable and the methods of quantitative genetics may help in doing this. At present the value of quantitative genetics lies in providing a base for planning experiments efficiently in the first stages of selection programmes.

4

Hybridization between Species and Races

Species hybridization is a very interesting subject. The successful hybridizer who produces a previously unknown combination feels that he has contributed something new, which in all likelihood would not have been produced in nature. And there is a very good chance that this new combination is also useful. Many of our most important crop plants are of hybrid origin, and new hybrid trees have definite promise.

A complete hybridization study for a single genus is a lifetime undertaking. Work by taxonomists on rare natural crosses or on introgression will provide clues as to which crosses can be made; for example, natural hybrids indicate the crossability pattern in the genus *Eucalyptus*. These clues should be supplemented by a good deal of controlled pollination work to develop a satisfactory crossability pattern because many of the most easily crossed species do not occur together in nature.

There is still much work to be done after the crossability studies, which will show that species A and B can be crossed and perhaps that the hybrids are promising. How can the hybrids be mass-produced, and which particular races or individual trees of the parent species should be used to give the best results? Those are questions the answers to which may need many years of research.

The role of hybridization varies among regions and among genera. There are groups such as *Castanea* in which the best timber species are deficient in a single character (in this case, disease resistance), which can best be obtained by crossing with another species. The majority of hybrids are intermediate between their parents in morphology and site requirements. Thus hybrids may be easily made but be of little value to forestry when large areas of intermediate habitats are not available. Sometimes disturbance by man has created the intermediate habitats on which intermediate trees are best adapted, showing hybrid vigour, and there are examples of this in *Eucalyptus* and *Aesculus*. Again, a species sometimes finds its largest use outside its natural range, in which case hybrids may prove better adapted, an example being the hybrid *Larix decidua* x *leptolepis*.

Hybridization offers relatively little hope of improving the American *Pinus resinosa*, which has resisted nearly all attempts to cross it with other species. Id the southeastern part of the United States the Southern pines are unlikely to be improved by interspecific hybridization; they reproduce readily and grow fast, while there are no other species of *Pinus* which surpass them in wood specific gravity - one of their most important properties. In any case, hybrids must be tested empirically before their potentials can be realized.

NATURAL HYBRIDS AND INTROGRESSION

Reports of natural hybrids between species belonging to the same genus are frequent. Also, there is one well-authenticated report of hybrids between tree genera, namely the cross *Cupressus* x *Chamaecyparis*. Evidence on the presumed *Tsuga* x *Picea* hybrid is less convincing.

The probability of obtaining natural hybrids is great when one sows seed collected from isolated specimens growing in an arboretum or park. If a species does not ordinarily self-pollinate under such conditions, most of the seed produced may be from crossing with another species. Also, the nursery practice is usually intensive enough to permit recognition of unusual types. Most *Abies* and *Acer* natural hybrids have occurred under such circumstances. So have a large number of those reported in *Picea, Pinus Populus* and *Quercus*.

In the majority of cases, natural hybridization leads to the production of one or a few F_1 trees only. The group of hybrids is in existence for one generation, then disappears. The next occurrence of the same combination may be in a quite different locality. This is true for most American reports of *Quercus* hybrids. It is also true of some interesting pine hybrids, such as *Pinus sylvestris* x *nigra* and *P. palustris* x *taeda*.

In *Populus, Pinus* and *Picea* study of the once-occurring natural hybrids provided the initial impetus for the later intensive work on artificial hybridization, but it was the study of the thousands of artificial hybrids which provided most information. On the other hand, study of the once-occurring hybrids of *Quercus* and *Aesculus* has contributed most of the knowledge about hybridization in those genera because artificial hybrids would be difficult to produce in quantity.

Natural hybrid swarms are infrequent but instructive. They occur most commonly on "hybrid" habitats, especially on disturbed areas. The oaks of southwestern United States furnish some of the best-studied examples (Muller, 1952). In central and northern *Texas Quercus havardii* and *Q. stellata* typically occupy sandy and clay soils respectively. Hybrids are common at the zone of contact between the two soils but not away from the zone of contact. The "pure" form of the species is found on the "pure" habitats typical of that species.

In the same region *Q. mohriana* and *Q. havardii* occur on soils derived from limestone and sand respectively. Hybrids are found where the soil is derived from a mixture of limestone and sand but not on the "pure" habitats. In California the hybrid species *Quercus* x *alvordiana* is really a group of hybrid swarms between *Q. douglasii* and *Q. turbinella* and is found only on intermediate habitats where the two parents come in close contact. Some of these *Quercus* x *alvordiana* swarms have persisted for many generations since the close of the Pleistocene Period.

In Australia *Eucalyptus dalrympleana* was one of the components of the original forest on undisturbed sites. As a result of man's activities, many of those sites have been damaged badly by fire. Now such areas are being invaded by hybrid swarms of *E. pauciflora* x *dives* (Pryor, 1957) whereas under natural conditions the two species excluded each other because of competition. Several other examples of hybrid swarms of *Eucalyptus* have recently been found on disturbed habitats.

Aesculus sect. *Pavia* includes four species native to the central and southeastern United States. In the original forest the species remained distinct. If hybrids were formed they occurred singly or in very small groups. Now there are many road cuts and cut-over areas on which the hybrids outgrow the parents. On such intermediate habitats there are occasional hybrid swarms of *A. glabra* x *pavia, A. glabra* x *octandra,* and *A. octandra* x *sylvatica.* In the latter two cases hybridization has led to considerable gene flow from *A. octandra* to the other two species. In yet another combination, *A. pavia* x *sylvatica* there is considerable gene flow in both directions from one parental species to the other.

By introgression is meant an extended period of hybridization followed by a considerable flow of genes from one species to another. Frequently it is hard to distinguish between the coming together of species by introgression and the divergence of species or races. In *Picea* introgression is suspected of being a major factor contributing to the wide adaptability of the northern, circumpolar species although it was proved in only two cases (Wright, 1955); *Picea engelmannii* and *P. glauca* cross readily in interior British Columbia and Alberta. Evidence of one species is found in the other far from the zone of contact.

The spruce population of northern Scandinavia and adjacent parts of the U.S.S.R. is a mixture of intergrades between *P. abies* and *P. obovata*. In other cases widely distributed species with adjacent ranges are known to cross artificially and are presumed to have crossed naturally when climate favored a range overlap. On the other hand, lack of genetic variability and lack of recent opportunity for hybridization are characteristic of the widely scattered southern *Picea* species.

Following the introduction of Populus deltoides from North America into southern Europe there has been a large amount of introgression from that species into *P. nigra*. Pure *P. nigra* is now rare in many localities and planting stock consists mostly of hybrids (Schreiner, 1959). True introgression is rare in *Quercus* despite the large number of species which have been reported.

Intermediates between the north European species *Quercus robur* and *Q. petraea* occur where their ranges overlap and may be hybrids. In southwestern United States *Quercus* x *drummondii* seems to be an introgressive derivative of *Q. stellata* with an infusion of *Q. gambelii* genes. In most other cases the species have maintained their identities even when growing in mixture with several other oak species (Palmer, 1948).

Three factors help to explain why hybridization sometimes leads to only a few intermediates and sometimes to hybrid swarms or introgression. First, there may be limited opportunity for crossing.

A few natural hybrids have been reported from arboreta for the cross between *P. strobus* from eastern America and *P. griffithii* from the Himalayas. These hybrids frequently appear in northern Italy where the two species are often grown together in gardens and parks. The cross is easily made artificially and introgression might be expected if the parents had the same range. Differences in flowering time may be all that keep *Acer rubrum* and *A. saccharinum* from becoming one species. Natural hybrids have not been reported although the two species commonly grow together and can be crossed with ease.

Secondly, the hybrids may be sterile. As yet there are few instances in which this is known to be the case.

The third factor is the nature and amount of selection pressure. Suppose that genes A and a are allelomorphic to each other, with respective frequencies of p =.9 and q =.1. Under random mating the zygotes will be present with relative frequencies of $(p + q)^2 = p^2 + 2pq + q^2 = .81AA + .18Aa + .01aa$. If selection favours the heterozygotes for one generation, the proportion of *Aa* trees will increase to a maximum (the theoretical maximum being.50).

But if environmental conditions then change so that selection favours the homozygotes, the Aa trees will tend to disappear. With complete elimination of heterozygotes the hybrid group in which zygote frequencies were*.81AA:.18Aa:.01aa* will produce gametes (pollen and egg cells) in which the gene frequency is*.988A:.012a.* These gametes will produce a new generation of trees in which the zygote frequencies are*.9757AA:.0242Aa:.0001aa.* Notice that selection against the heterozygotes has reduced the frequency of *Aa* trees by about 90 per cent and reduced the frequency of the less common homozygote (*aa*) by about 99.9 per cent. In other words, very few a genes remain with which to form *Aa* trees in the following generation. If the direction of selection should again be reversed to favour *Aa* trees, the build up of a hybrid swarm would have to start afresh.

This mathematical example has two corollaries. First, the existence of a persistent hybrid swarm is strong evidence that selection favours heterozygotes or, in other words, that usable hybrid vigour is present. Second, the absence of a hybrid swarm does not necessarily mean that useful hybrids cannot be produced. Man can take advantage of hybrid vigour exhibited in two generations out of three whereas nature cannot.

If a breeder's interest is in the overdominance indicated by the presence of a hybrid swarm, he should attempt to mass-produce by artificial means F_1 or possibly F_2 populations because it is in those generations that the overdominance will be most pronounced. However, if he is not in a position to undertake such an intensive programme he should not attempt an intermediate measure such as a genetic thinning. By another theorem from population genetics, it can be shown that such a hybrid population will come to a natural equilibrium when the,

$$\text{frequency of the a gene} = q = \frac{s1}{s1 + s2}$$

s1 and s2 being the selection pressure against the *AA* and *aa* trees respectively. It can also be shown that the population is most productive at this natural equilibrium (if it cannot be maintained in a completely heterozygous state by artificially crossing *AA* and aa trees). As an example, if *AA*, *Aa*, and aa trees have fitness ratings of.60, 1.00, and.85 respectively, the hybrid population will have an average fitness of.89 if at equilibrium. Artificial selection to favour the best trees (which would be the heterozygotes) would increase the number of unproductive *AA* trees and would lower the average fitness to.862 whereas artificial pollination to ensure that only *aa* trees were planted would increase the average fitness to 1.00.

If a breeder's interest is primarily in additive genetic variance, a natural hybrid swarm forms an excellent starting point for a selection programme. In southern Michigan there is a large group of hybrids between *Pinus thunbergii* and *P. densiflora.* In several characters the additive genetic variance within it is probably two or three times that found in any natural population of the pure species. Heritability and response to selection would also be two or three times as great as in a population of the pure species.

CROSSABILITY PATTERNS

With the exception of *Cupressus* x *Chamaecyparis*, mentioned earlier in the chapter, intergeneric hybrids in temperate zone trees can be disregarded and attention can be focused on crosses within genera. The crossability patterns are best known in *Populus, Pinus Picea Quercus, Castanea,* and *Larix*. In those groups enough crosses have been attempted to formulate general rules which permit one to forecast which species combinations are most likely to be successful. For *Acer, Aesculus, Betula, Fraxinus* and *Ulmus*, information is more fragmentary and permits only the broadest generalizations. For most other genera (including all the tropical ones) it is even more fragmentary, consisting mostly of lists of natural hybrids.

The amount of genetic differentiation within what is considered to be a single genus varies greatly. At one extreme is *Populus*. There has been so little differentiation that a given species can apparently be crossed with any other species. *Castanea* may have a similar pattern - all attempted species crosses have been successful. Genetic differentiation is a little more pronounced in *Picea*. Most species can be crossed with their geographic neighbours but not with those species from a distant locality.

There is no sharp break into distinct crossability groups and it is possible (but this has not been tested) that any species could be linked to any other one by crossing with intermediate types. *Eucalyptus* exhibits a similar pattern, neighbouring species in the same section crossing with relative ease. *Quercus* is divided into 3 subgenera (only 2 are known genetically). A large number of hybrids between species belonging to the white oak subgenus *Lepidobalanus* have been reported and possibly any two white oaks could be hybridized. The same is true of the red oak subgenus *Erythrobalanus*. On the other hand, there are no convincing demonstrations of hybridization between the subgenera of *Quercus*.

The pines are at the other extreme. The genus *Pinus* is divided into 2 subgenera and a large number of series. With 3 exceptions attempts to cross members of the same series have been successful. The pattern in *Acer* is similar to that in the pines. Of 60 artificial crosses which have been attempted, 6 succeeded.

The successes were all within one of the 13 recognized sections. From the genetic standpoint, series or sections in these 2 genera have approximately the same amount of differentiation as some entire genera. It would make sense biologically (but not practically) to regard these small groups of species as separate genera, as some authors have done in the past.

Morphological or physiological similarity and geographic distribution provide excellent clues as to the crossability pattern in a genus. The most easily made combinations are between similar species with nearly overlapping ranges. Often distance was a sufficient barrier to render unnecessary another type of isolating mechanism. If this barrier is removed by interplanting or by controlled pollination, hybrids result. On the other hand, species which occur in mixture in the same stands usually do not cross.

Sometimes differences in flowering time or in microsite requirements provide an isolating barrier which can easily be overcome by artificial crossing. Sometimes the isolating barrier arises as the result of genic or chromosomal differentiation. Irradiation of pollen and the use of pollen mixtures offer some promise of overcoming such a barrier. It is also difficult to cross morphologically dissimilar species or species with widely separated distribution areas. In such cases, long separation has usually been accompanied by great genie or chromosomal differentiation.

Crosses which fail in one environment may succeed at another place or time. Knowledge of the nature of the barrier often helps to break it down.

Examples of the relationship between crossability, similarity, and geographic distribution are numerous, and for illustration the genus *Pinus* can be cited. The series *Strobi* comprises several similar species. All except one have separate (but neighbouring, if considered in the geological sense) distribution areas and can be crossed with each other rather easily. The single exception (*Pinus lambertiana*) is not typical of the series and does not cross with the other members. Several attempts have been made to cross species of ser. *Strobi* with members of other series but only one such attempt succeeded.

Pinus ser. *Sylvestres* comprises about a dozen hard pines. The Asiatic members are morphologically similar and can be crossed with each other in nearly all possible combinations. They are less similar to the European representatives of the series. Only a few European x Asiatic crosses successful and these gave low seed sets. The single United States representative in series *Sylvestres (P. resinosa)* has several distinctive morphological and chromosomal characters. Over 20 years of effort by many people resulted in only 4 hybrid trees between *P. resinosa* and another species.

Pinus ser. *Australes* comprises groups of southeastern and southwestern United States species. Members of each group are more similar to each other than to members of the other group. All possible southeastern x southeastern species combinations and a great many southwestern x southwestern combinations have resulted in hybrids. However, no southwestern x southeastern combination has been successful (Duffield, 1952).

The spruces (*Picea*) also show a relation between crossability, similarity, and distribution. In northern United States, Canada, and northeast Asia is a group of 6 species, each with a nearly separate distribution area and each with certain similarities to its neighbours. All attempted hybridizations within the group were successful but several attempts to cross those spruces with others outside it were not. Across northern Europe, Siberia, and northern China is another "chain" of similar species, within which crossing occurs very easily. The 6 species which inhabit the same mountain in Japan do not cross, but one of them can be crossed with mainland spruces. One of the notable exceptions to the general crossability pattern is the cross *P. omorika x sitchensis*, which has been successful in Europe.

The genus *Larix* contains about 10 species, each with its own separate geographic range. Apparently differentiation has been entirely geographic and the species cross easily.

Of 7 attempted artificial hybridizations in *Fraxinus* only one succeeded. That was between an eastern United States and a western United States representative of the same section. The 2 species can be considered neighbours in the geological sense and are so similar that one could mistake aberrant specimens of one for the other. Hybrid swarms are reported between members of a similar species-pair in southwestern Europe.

Several cases are known in which crosses between species with different chromosome numbers yielded good seed. The general relationship between chromosome number and crossability has not yet been worked out.

These generalizations can be summarized by pointing out that a good monograph and a detailed set of range maps are of great help in planning a species crossing programme. If it is assumed that there are 90 species in the genus *Pinus*, there would be a total of (90 x 89)/2 = 4,005 different species x species combinations. Work to date indicates that 10 per cent or 400 of these might be successful. A breeder who crossed at random would have only a 10 per cent probability of success. But one who studied the taxonomy and distribution could increase this probability to 80 per cent.

BREEDING FOR HYBRID VIGOUR IN THE F1 AND F2 GENERATIONS

Most emphasis in species hybridization has been on the production of vigorous F_1 hybrids. Enough work has been done to show that the approach is extremely promising. The majority of the hybrids are useless but enough possess hybrid vigour to make the work worthwhile. Often hybrid vigour is evident in one part but not in another part of the range of a species.

The introduction of a successful F_1 hybrid combination into commercial practice is carried out in four stages. First comes a crossability study to determine what combinations can be made. As already indicated, this work is well advanced in some genera and scarcely started in others. Next comes the production of a few hybrids of each promising combination and their testing in comparison with the non-hybrids they might replace. A good start has been made in this direction, but in scarcely any case have the tests been extensive enough to indicate a hybrid's true potentialities. Most of the *Pinus* hybrids produced in California and Pennsylvania are scarcely known outside a few test localities; and the potentialities of *Larix decidua x leptolepis*, which is very useful in Europe, are scarcely known in America. The third stage is the determination of mass production possibilities.

This is no problem in those species of *Populus* which can be propagated readily by cuttings, however, it is a very real problem in other genera. The Institute of Forest Genetics at Placerville, California, did enough work on several combinations of *Pinus* to show that mass production is possible. In recent years, the Korean Institute of Forest Genetics has done a large amount of work, and obtained extremely promising results, on a few *Pinus* combinations. The fourth step in introducing a successful F_1 hybrid into commercial practice is determining which particular trees would make the best parents.

This has barely started for any species combination. Take the cross *Pinus sylvestris x densiflora* for example. It was produced in the United States, originally in Connecticut and later in Pennsylvania. In both cases the first available trees, of unknown origin, were used as parents. The hybrids grew extremely fast but cannot be used because of poor form and low seed set.

The Pacific southwest and the northeastern forest experiment stations of the United States Forest Service have produced a large number of species combinations in *Pinus* but only a few can be mentioned here. *Pinus monticola x strobus* is one of the most promising. Seed sets are high enough to indicate good mass production possibilities; growth in California and in many other parts of the United States equals or exceeds that of the faster-growing parent; there is intensive work under way on breeding disease-resistant cultivars of both parent species. When the work is completed, truly remarkable F_1 hybrids will be possible. The hybrid *P. attenuata x radiata* also shows promise. It could be mass-produced. Although it does not outgrow the faster-growing parent *(P. radiata)* on warm sites, it promises to be one of the most productive pines on sites too cold for that species.

Seed sets are very high for the hybrid *P. densiflora x thunbergii*. This combination shows considerable hybrid vigour in Pennsylvania, but not in Japan where the parents are native. The potential usefulness of a fourth combination, *P. nigra x resinosa*, is limited because of mass production difficulties. Only four hybrids were produced after many years' effort. Thus, these hybrids have good growth characters but remain botanical curiosities.

The potentialities of *F. pine* hybrids are being exploited to the full in Korea. There thousands of bags are used to make the cross *P. rigida x taeda* each year and the programme is proving to be economic. The hybrids can produce the most income even though they are more expensive to produce than non-hybrids.

The cross *Larix decidua x leptolepis* is the subject of more than 27 different publications from as many parts of northern Europe, almost all reporting greater growth than for either parent species (for the references see Wright, 1962). Seed sets are high but *Larix* does not lend itself as well to mass-controlled pollinations as does *Pinus* Syrach Larsen of Denmark has therefore designed a seed orchard containing a self-incompatible clone of one species and seedlings or clones of the other. The self-incompatible clone serves as the female parent and produces hybrid seed.

A small amount of American and a much greater amount of Russian work has resulted in the production of several artificial hybrid combinations in *Quercus*. Some of these are intermediate between their parents but others exhibit useful hybrid vigour (Piatnitsky, 1960). However, at present even the latter appear to be of very limited usefulness. Controlled pollination is extremely tedious and likely to be too expensive to be done on a commercial scale. Vegetative propagation is impractical. Seed orchards where natural crossing can take place may be the answer, but efforts to find trees which set large percentages of hybrid seed after natural gross pollination have been unrewarding.

The silviculture of *Populus* is now based much more on hybrids than on pure species. Cultivars derived from crosses involving *P. deltoides, P. nigra P. trichocarpa* and other species are commonly planted in Europe. Most of these have not been compared with their parents but hybrid vigour is usually assumed. As the result of Schreiner's pioneer work in the northeastern United States, many promising hybrids have been tested in America and can soon be planted on a commercial scale.

Species of *Populus* other than the aspens can be propagated readily by cuttings so that mass production is no problem. Ease of rooting has also permitted breeders to conduct a large number of tests for clonal differences within F_1 full-sib families. As expected, significant clonal differences were rarely found except when the parents were themselves hybrids.

In short-term tests *Acer platanoides x mayrii* and *A. platanoides x cappadocicum* exhibited some hybrid vigour. The crosses were easily made on an experimental basis at the rate of a few hundred per day. However, that rate could not be increased greatly because of the flower structure, and mass-controlled pollination does not seem feasible. Neither do seed orchards because of differences in flowering time. Polyploidy is possibly the answer. Perhaps the few hybrids now available could be made truebreeding by having their chromosome number doubled. But would the tetraploids be as desirable as the diploids? That must be determined.

Theoretically one can expect a certain amount of hybrid vigour to persist into the F_2 generation. This raises the possibility of utilizing the F_2 generation if the F_1 cannot be produced cheaply. At least in some of the pines the F_1's are sufficiently fertile for F_2 seed to be produced in quantity. The Institute of Forest Genetics in California has been pursuing this possibility for several Pinus, combinations. In the United Kingdom there are plans to raise large F_2 progenies of European x Japanese larch *(Larix decidua* x *leptolepis)* in the expectation that some hybrid vigour will persist.

In most of the illustrations quoted here, the hybrids are still young, since there are no adequate older tests of many of the most interesting combinations. Also, many of the preliminary growth tests have been conducted on only one or two sites, and it is well known that hybrid vigour may be evident at one place and not at another. There is, therefore, considerable testing still to be done.

Practically all the *Pinus, Picea* and *Populus* hybrids to date have been made with the first available parents. This leaves considerable work for future tree breeders. The first *Pinus, densiflora x thunbergii* hybrids produced in Philadelphia seem quite useful but it is inconceivable that they can be the best.

It is much more likely that even better hybrids can be produced by crossing selected races of the two parent species or perhaps selected individuals within those selected races.

POSSIBILITIES OF RECOMBINATION IN LATER GENERATIONS

Almost every tree hybridizer hopes he will produce an F_1 or F_2 generation which possesses enough hybrid vigour to be worth planting on a large scale. That is not always possible. Sometimes segregates arising in a later generation will be the valuable product. This will be true under three conditions:

1. If controlled pollination is so difficult that F_1's and F_2's cannot be mass-produced, (as in *Juglans, Castanea, Acer, Quercus* and possibly *Abies*);
2. If hybrid vigour is present but is not sufficient to make a plantable cultivar *(Castanea dentata x mollissima);*
3. If the genetic variance is mostly additive.

If it is desired to incorporate one character from species A with many characters from species B, back-crossing is indicated. The F_1 hybrid between A and B should be produced artificially, then backcrossed to species B and the process repeated again if necessary. In crop plants four or five generations of backcrossing are frequently necessary. If it is desired to incorporate approximately equal amounts of species A and B, an F_1 generation is produced; the F_1's are crossed with each other to produce an F_2, those are crossed with each other to make an F_3, etc.

The necessity for this approach has been appreciated for many years but investigators have been loath to follow it because of the long periods of time involved. Such delays may be necessary but need not be interminable. Numerous, important tree species produce flowers at ages from one to five years. If a tree breeder had known their desirability at the start, within 20 years he could have had F_4 generations in production.

At present most many-generation breeding projects are unplanned developments of F_1 hybridization studies. This situation must be changed if the most rapid progress is to be made. The field tests designed to produce a later generation segregate should be different from those designed to test the usefulness of the F_1 or F_2 generation *per se*.

The aim in a many-generation project should be to produce and test an F_3 or second generation backcross as rapidly as possible. Why? The F_1 generation is a uniform one, offering almost no opportunity for selection. [2] Suppose that there are 25 F_1 seedlings and 10 of them fruit at a very early age. The genetic component of the variation in the F_1 population is almost nil, so that it makes little difference whether those 10 are used as parents of an F_2 population or there is a wait of several years in order to see whether the other 15 might have superior phenotypes. The F_2 population is a genetically variable one and offers opportunity for genetic selection. But a large proportion of the F_2 genetic variance is non-additive and therefore cannot be fixed by crossing the best trees. This means that a breeder is justified in getting only a moderately good evaluation of the F_2 because no matter how long he looks at the trees he cannot tell their breeding potential as well as if he were to wait until the F_3 or F_4.[3] It is better to look at the F_2 for relatively few years and to practice a mild selection. The 10 or 20 per cent best trees might be chosen and used as parents of a replicated F_3 test in which the identity of all families is maintained.

It has already been mentioned that selection is more productive in the F_3 than in the F_2 generation because of the smaller amount of non-additive genetic variance. Another reason for preferring the F_3 is that it offers an opportunity to practice family selection. This means selecting breeding stock on the basis of a tree's phenotype as well as on the performance of other trees having the same parent(s). This is much more effective than selection on the basis of phenotype alone.

Should one use control-pollinated (both parents known) or open-pollinated (only the seed parent known) progenies when conducting a selective breeding programme such as outlined here? The F_1 and F_2 populations can be relatively small, consisting of a few dozen or few hundred trees. They could well be control-pollinated. As long as a tree breeder is working with multiple factor inheritance (which is most likely), the F_3 population should be very large, containing scores or hundreds of progenies and thousands of individual seedlings.

The likelihood of success with small F_3 tests is too small to warrant the trouble involved. In *Populus Picea*, and many *Pinus* species controlled pollination can be easily done on the required scale. On the other hand, workers with genera in which controlled pollination is laborious, as in *Abies* in which the female flowers are rather inaccessible, or *Castanea Juglans* or *Acer* in which each pollination yields few seeds, would be well advised to work with open-pollinated progenies in order to have tests of adequate size.

Vegetative propagation is commercially possible with many species of *Populus*. Hence, in that genus it may be desirable to establish clonal tests in the F_2 or F_3 generations in order to select clones with desirable properties. In genera such as *Pinus* and *Quercus*, clonal tests are expensive and probably not needed because seedling families will be planted commercially.

Family selection in segregating generations will play a large role in the *Castanea* breeding programme. The American *C. dentata* is a fast-growing timber tree with desirable growth habit and wood properties but without resistance to chestnut blight *(Endothia parasitica)*. That must be furnished by the Asiatic *C. mollissima* or *C. crenata*, neither of which becomes a large timber tree. A moderate number of F_1 hybrids have been produced which show more or less intermediacy between the parents. Also, moderate numbers of F_2's and backcrosses have been produced. Some of the F_2 segregates have good vigour and much more resistance than the *C. dentata* parent. That is not enough, however, and it appears that large F_3 and F_4 generations must be raised to yield a cultivar which is commercially acceptable for timber production.

A great deal of the *Populus* breeding work in southern Europe is probably aimed towards the production of desirable later-generation segregates. However, no formal many-generation programme has been started, selection being among trees of unknown generation.

Several F_1 *Pinus* hybrids have reached fruiting size at the Institute of Forest Genetics in California. Most have been fertile. One of the most promising crosses is *P. jeffreyi* x *coulteri.* Some F_2's and backcrosses exhibit the desirable growth characters of the former and resistance to the pine reproduction weevil *(Cylindrocopturus eatoni)* of the latter. The degree of recombination is sufficient to indicate that a group of F_3 families might comprise a commercially acceptable cultivar.

INTERRACIAL HYBRIDIZATION

The crossing of different races within a species offers the same possibilities as does the crossing of different cultivars of a crop plant or different breeds of animals. The F_1 hybrids may exhibit hybrid vigour as do crosses between inbred lines of maize. Or they may exhibit a combination of desirable characters and furnish the foundation for a several-generation selection programme.

The pioneer German tree breeder, Dengler, made a start on interracial hybridization in 1926 when he crossed German, French, and Scottish origins of *Pinus sylvestris* He thought some combinations showed hybrid vigour at 10 years old, but Scamoni later reported, when the trees were 20 years old, that the hybrids were intermediate between their parents.

In 1948 Johnsson and his co-workers at Ekebo, Sweden, made a systematic series of crosses among European aspens *(Populus tremula)* originating at three different latitudes in Sweden. At 10 years all the hybrids were considerably superior to the midparent mean and one combination (southern x central) was taller than the faster-growing parent.

Other Swedish tree breeders, especially Nilsson and Langlet, have had promising results from interracial hybridization in *Pinus sylvestris* and *Picea abies*. These were seen during the study tours preceding the world consultation. Following their lead, workers in other regions will undoubtedly show greatly increased interest in this subject.

FACILITIES NEEDED FOR HYBRIDIZATION WORK

Good collections of breeding material are the prime requisite for productive hybridization work. The breeder interested in determining a crossability pattern needs several flowering specimens of several races or species. The breeder interested in mass-producing a particularly promising hybrid needs many selected specimens of each parent. One or two specimens of each parental type are not enough because they may not fruit nor give enough non-hybrid seed to provide controls.

In temperate climates a period in the spring should be reserved for at least five years when embarking on a serious hybridization programme. Time is needed for bagging, pollinating, removal of bags and observation of flower development. Seed sets are often so variable that the same cross needs to be repeated a few times to establish a reliable average pattern.

A hybridizer needs a small but good research nursery and several test sites. Usually few seeds are available of combinations which are made for the first time. This means that the testing must be done with greater precision than is usual in planting research. Both nursery and experimental forests must, therefore, be equipped to give the tests better than average care.

If the breeder is interested in developing both F_1 and F_3 cultivars, duplicate test plantings may be necessary. The design used to determine comparative growth rates of the F_1's and their parents may not be the one which would insure earliest fruit production. Also, a good design to test hybrid vigour would not necessarily be the one used to obtain F_2 and F_3 hybrids by natural open pollination.

5

Forest Hydrology and Management

INTRODUCTION

Forest hydrology has built a strong foundation of general principles concerning the direct effects of forest management on hydrologic processes from plot studies, process studies, and watershed experiments. The challenge now is to apply these principles to predict how hydrologic processes will respond to many forms of change in forest landscapes. Forest hydrologists have long recognized the need to understand indirect and interacting effects of forest management at much larger spatial scales and longer temporal scales than is possible in plot studies, process studies, and watershed experiments. Indirect effects are responses to forest management that are displaced in time or space, such as fire suppression leading to insect outbreaks that affect forest hydrology. Interacting effects occur when two or more management practices coincide, such as when post-salvage logging and road building have a different collective effect on forest hydrology that differs from their individual effects.

This chapter examines the research challenges faced by forest hydrology as it moves from principles to prediction at larger spatial scales, at longer temporal scales, and in a changing social context. The chapter concludes by outlining the potential for improved cumulative watershed effects analysis that could provide the predictions needed by forest and water managers in the twenty-first century.

SPATIAL RESEARCH NEEDS

A key unresolved issue in forest hydrology is how to apply the findings of hydrological studies in one area to a different area or how to scale up the findings to large watersheds and landscapes. Although forest hydrologists have confidence in the general principles of hydrologic responses to forest management and disturbance, they cannot predict precisely how forest management will affect hydrologic processes in specific places other than those that have been intensively studied. Most forest hydrology studies are conducted in small watersheds that are instrumented to measure streamflow and other hydrologic properties. However, the sum total of the area studied by forest hydrology is only a tiny fraction of the watersheds in the United States, and hydrologists recognize the need to extend hydrologic knowledge from "gauged basins" (watersheds that have measured records) to ungauged basins. Predictions are most needed in ungauged places to better understand hydrologic effects where conflicts sometime arise: in water supply systems for agriculture and cities; rivers where endangered and threatened aquatic species occur; large water bodies such as the Chesapeake Bay; and many, many others.

Forest hydrology is adopting a landscape perspective to examine spatial patterns of forests and associated hydrologic processes and to link principles from plot- and small watershed scales to predictions at larger spatial scales (hundreds of square kilometers). Within a watershed, forests can be located in the headwaters, along riparian corridors, in woodlots in agriculture lands, and in urban or suburban areas. Based on its intra-basin position, a forest fulfills various water-related functions with respect to water quantity and quality. For example, forests in headwaters influence water yield and the quality of water delivered to downstream areas. Riparian forests located along streams throughout a watershed provide key functions for protecting streams from inputs of sediment, nutrients, and herbicides; provide wildlife habitat for terrestrial and aquatic organisms; and support a diversity of other functions. Riparian forests have been greatly altered by economic development, and they are the focus of many forest management guidelines designed to preserve water quantity and water quality.

Hydrologists use models to predict water quantity and quality in watersheds where there are no measured records. Since 2004, a working group for Prediction in Ungauged Basins (PUB) of the International Association of Hydrological Sciences has developed methodologies for assessing uncertainty in hydrologic predictions arising from uncertainties in landscape properties and climate inputs, choice of model structure, and methods of information transfer from gauged to ungauged watersheds. Most hydrological models are developed and tested for gauged basins and subsequently are validated and applied to ungauged areas.

However, models that have been fitted to data in small, gauged watersheds often provide inaccurate or imprecise predictions when they are (1) extrapolated to other small forested headwater basins, (2) extrapolated to future time periods, or (3) applied to large watersheds. This problem of prediction in ungauged basins has preoccupied hydrology researchers for several decades, and is compounded by a lack of information about how direct hydrologic effects interact under the multiple sets of specific conditions that occur in changing forest landscapes. By examining forest hydrologic processes under a wide range of conditions, landscape-scale studies could provide data and understanding to help extend basic forest hydrology principles to make predictions needed by water managers.

Road networks are a pervasive feature of forest landscapes. The location and density of roads in a watershed can influence the hydrologic effects of roads. In many forested areas, roads are concentrated in the valley bottoms immediately adjacent to streams, meadows, and wetlands. These roads have a particularly high potential for delivering run-off and sediment to streams, lakes, and aquatic ecosystems. The legacy of midslope roads from past logging practices is also of concern, because these have a high potential for subsurface flow interception, connectivity to the drainage network, and initiating shallow landslides. Considerable research effort has been devoted to modifying road design and management to mitigate erosion, and to developing techniques to decommission roads. The hydrologic effects of road networks at large scales, and the effects of road decommissioning are not widely studied.

TEMPORAL RESEARCH NEEDS

Water management systems have been designed and operated under the assumption that hydrologic variables such as annual water yield, while varying over time, can be predicted reliably based on instrument records. However, increased understanding of long-term variability and trends in climate has undermined this assumption. If precipitation and streamflow vary over the long term, future annual water yields may fall short of the levels that water supply systems—and attendant agricultural and urban development—were designed to provide. Given this context, it is critical for forest hydrologists to extend beyond general principles to make predictions of how forest management and disturbance affect hydrologic response on time scales that exceed those of most forest hydrology science.

Some forest and stream management plans now include the historical range of variability, which presupposes that (1) past conditions and processes provide context and guidance for managing ecological systems today; and (2) disturbance-driven spatial and temporal variability is a vital attribute of nearly all ecological systems. The historical range of variability helps characterize the variation in quantity, quality, and timing of streamflow from forests. It can also be used to establish baselines for assessing change in water from forests over time.

Forests and their associated hydrologic processes change on time scales ranging from decades to hundreds, or even thousands, of years. The temporal context for understanding forest hydrologic processes involves expanding the temporal scale into the past to consider the effects of past forest practices and into the future to project and anticipate changes in land use and climate. Many different kinds of legacies of past human activities affect forests in the United States. In some areas, native forests have been converted to agricultural and urban uses, and forests have regrown on abandoned agricultural lands in others. Roads and expansion of urban areas have fragmented forests into smaller, less-contiguous patches and created new drainage patterns. Fire suppression has changed the structure and community composition of many forests, especially in those with otherwise active fire regimes. Exotic species introductions, grazing by domestic animals, predator eradication, and timber harvesting methods have changed forest cover and species composition as well. Future urban and suburban development and climate change are expected to continue to alter forest cover and species composition. Human activities will modify forests in the future, and future legacies will reflect current, regional forest histories.

THE HYDROLOGICAL CYCLE OR THE WATER ECONOMY

Forest ecosystems require water which is the most important input for their survival. in particular, under tropical monsoon climatic conditions, forest ecosystems play a vital role in moderating the impact of rainfall and controlling the instant run-off of water.

The impact of forest ecosystems on rainfall has been a topic of popular debate. Meher-Homji (1986) has pointed out that forest ecosystems play an important role in pre- and post-monsoon rain fall. This may not alter the amount of total rainfall significantly, but by providing protective soil moisturisation during a period significant for plant and growth, it plays a very crucial economic role.

The hydrological cycle describes the ecological processes involved after a drop of water has entered the forest ecosystem as rainfall or dew or even snow. In the meteorological conditions prevailing in the forests of India, except parts of Western Himalayas, the most significant form of water input to forest ecosystems is through rainfall. The hydrological cycle represents mainly the physical aspect s of essential ecological processes of a forest ecosystem. The hydrological cycle is an instrument for a fundamental understanding of forest ecology. On its stability depends the stability of the forest ecosystem. The source of all water required for the survival of plants, birds, animals and human population is precipitation (P) from the atmosphere. Once precipitation takes place, as rain, dew, snow, etc., water enters the forest ecosystem and is first intercepted by the forest canopy. Some amount of the incoming water evaporates back to the atmosphere in the process and does not touch the soil. This is known as interception loss. Some amount of water falling on the canopy flows down to the top soil as stemflow and some falls directly as throughfall. Some amount of water drips down after a delay period and is known as drip.

Of the total amount of water reaching the top soil, some flows out of the forest ecosystem as run-off (R) and is lost to the plant. The rest infiltrates to the soil and percolates to the underground aquifers to recharge the springs (1). Infiltration is encouraged in forest soils with a good cover of litter and a low density spongy humus. Compaction of the top soil by cattle or human intervention greatly enhances run-off and reduces infiltration.

The infiltrating water is first absorbed by the soil which holds moisture (M) in small pores through capillary action, and this capacity of the soil to hohl water against gravity is known as field capacity. When water availability exceeds field capacity it flows down under gravity and reaches the rock system to recharge the underground aquifers. The aquifers recharge the outflows through springs, and on forest slopes saturated soils give rise to seepage streams, which together with the surface run-offs join to form rivers. The moisture retained in the soil goes back to the atmosphere either as direct evaporation or as transpiration through green plants (ETR). Thus, the hydrological cycle leads to the water balance equation:

$$P = R + I + M + ETR$$

Where, P = Precipitation, R = Run-off, I = Infiltration Percolation, M-Soil Moisture Change, ETR = Evapotranspiration.

The relative amounts of.water flowing through the various routes in the cycle are influenced by the state of the canopy, the state of the ground cover and humus, type of soil, etc. The management of forest ecosystems will thus depend on the main economic objectives that the water output of forests are to satisfy. In the temperate regions of the world where precipitation is well distributed and in many parts the ground is covered by snow for a few months in a year and slopes are gentle, complete denudation of the catchment forests is recommended as a method of maximising water yield. On the contrary, in tropical and monsoon climate, forest ecosystems play a vital role in reducing run-off and encouraging infiltration through leaf litter and humus formation, thus ensuring a stable water yield. The management objective of a forest ecosystem will thus depend on the meteorological conditions and the manner in which the water economy is to be developed because under certain conditions water and not biomass is the most important economic output of forests.

THE NUTRIENT CYCLE AND THE SOIL ECONOMY

The flow of water in the forest ecosystem plays the vital role of carrying the nutrients required for plant growth from the soil and controls the rate of uptake of nutrients. The nutrient cycle thus represents the chemical aspects of ecological processes of a forest ecosystem. In economic terms, the nutrient cycle describes the economy of the soil, describing and quantifying the nutrient uptake from and return.to the soil on which the forest grows.

The botanical process of plant growth requires a large number of elements like hydrogen, carbon and oxygen, macro-nutrients like Calcium, Potassium, Magnesium, Nitrogen, Sulphur and Phosphorus, and micro-nutrients like Boron, Copper, Iron, Manganese, Zinc and Molybdenum. Elements like CH and O are available from water and Carbondioxide from the atmosphere.

These nutrients with the exception of N are available from the weathering of rock mineral and Nitrogen is available from the atmosphere. Apart from the biomass exported out of the forest ecosystem, these nutrients eventually return to the soil surface through deadwood and litter fall as well as washing of foliage by rain water. On the soil surface a variety of forest floor fauna including micro-organisms and bacteria transform the biomass through decomposition and release nutrients for further plant nutrition. The uptake and return of nutrients in the forest ecosystem is well studied in the temperate regions of Europe or the USA. Unfortunately, the soil economy associated with indigenous tree species in tropical countries like India is least understood. This obviously leads to wide gaps in knowledge for the proper choice of species and their management in afforestation programmes.

In every forest ecosystem the nutrients that are used by the trees are normally returned to the soil completely. When forest biomass is extracted and transported for the satisfaction of sustenance needs or industrial/commercial demands, substantial amounts of nutrients go out of the forest ecosystem, and for intensive forest exploitation artificial fertilisation of the forest soil becomes essential.

Nutrients are supplied to the forest trees both from the atmosphere and the soil. Nitrogen is available from the atmosphere directly, as dissolved nutrients in the rain water and as particulate deposition which gets washed down to the forest floor through rainfall. Rain and wind erosion transport the nutrients from the parent rocks to the soil and the moisture in the soil dissolves them and transfers them to the body of the trees.

The nutrients are returned to the soil through the litter which contains organic remains of plants, like leaves, barks and twigs in exploited forests and organic remains of animals on the soil surface or in the top soil layer. In tropical rain forests leaf litter is about 10 tons per ha while in the open conifer forests it may be only I ton per ha. One part of green plants combine with the litter as animal waste through the consumption of green matter by the herbivores.

The accumulated leaf fall and other forms of litter then begin to decompose on the forest floor through the action of micro-organisms present in the soil. In tropical conditions where soil biotic activity is encouraged by relatively higher temperatures, the rate of decomposition is quite rapid. Due to the content of the leaves and the soil chemistry, the rate of decomposition of litter in rapidly growing tropical forests is several times greater than in the conifer forests in the temperate regions.

The soil organisms that decompose the litter are mainly bacteria and they multiply in soil with earthworms. The soft parts of the plant are normally decomposed by micro-organisms alone but woody biomass are broken. down by a complex interaction, thus the return of the nutrients back to the soil is an intricate process involving many actors. As decomposition proceeds, the nutrients are released in the form of soluble ions that can be directly absorbed by the root system and the cycle starts once again.

The nutrient cycle is disturbed by the destabilisationof the hydrological cycle. With the opening up of the forest canopy and instant surface run-off increasing, the leaching of the nutrients increases and the nutrients available for new plant growth become less, thus setting in motion a process of decay in the forest ecosystem. In extreme cases of nutrient loss and continued exploitation of the forest biomass, the vegetational evolution is reversed and a full canopy forest gets degraded to scrub forest or grasslands.

Both the water economy and nutrient economy constitute nature's economy in forest ecosystems. They need to be stable in order to sustain the productivity of forest ecosystems. The two other economies, *i.e.*, the survival economy of the basic needs satisfaction of the people and the market economy of forest product demand of the industrial/commercial sector compete for the same resource base, forest biomass and generate conflicts over forest resources between the needs of nature and the people on the one hand; and between the needs of nature as well as people and market demand, on the other.

In the context of the forests of India, nature's economy and survival economy have always been overlapping and were simultaneously functioning without major conflicts as the small survival needs of the people were satisfied through a conservation-oriented utilisation managed by an informal but strict code of conduct towards forests.

It is thus reasonable to assume in the context of forest resource utilisation in the precolonial periods, that the satisfaction of survival needs was an intrinsic part of the functioning of forest ecosystems. This was particularly so because human settlements in India grew as an integral part of the forest ecosystem and not at the cost of it as was the case in industrialized countries in the last few centuries.

With the introduction of large-scale commercial exploitation of forests by the British this situation underwent a drastic change. A schematic picture of the three competing biomass requirements of the three economies. The horizontal axis represents the distance (D) from the core of the forest ecosystem while the vertical axis represents the quantity (Q) of biomass required by the three competing economies. Nature's requirement (ON) is spread throughout the forest ecosystem while the survival requirement (QS) is divided between inside and outside the forest ecosystem. It should be noted that the spread is not too far away 'from the forest ecosystem's boundary since only the local population can collect the' forest biomass.

The requirement of the market economy (QM) is high as well as spread over long distances far away from the forest ecosystem, since it can be transported over long distances. This indicates a continuous long distance transfer of large quantities of forest biomass outside the forest ecosystem. All forest related conflicts are thus based on conflicts between the above mentioned requirements-ON, QS, and QM. The objectives of forest management decide as to what should be the actual biomass quantities allocated to these diverse requirements. When forests are viewed as a complete ecosystem and not as a mechanical collection of wood producing trees, the management strategy of forests has to evolve along the ecosystems concept. The ecosystems approach has the objective of ensuring sustainable production of an optimum biomass mix so as to satisfy the demands of nature's economy and survival needs and produce biomass for commercial/industrial purposes to the extent possible. Accordingly, such an approach should

1. Lead to a harmonious utilisation of resources not resulting in external diseconomies (exploitation of forests for timber should not lead to floods and silting).
2. Provide for the equitable returns to various sections of society depending on the same resource system (exploitation of forests for timber should not conflict with the domestic requirements of fodder or wood as long as alternatives are not made available at a low cost).
3. Wherever externalities cannot be eliminated, the logic of costs and returns needs to be worked out for proper methods of compensation to those who bear the costs generated by activities that benefit others.

CONVERTING NATURAL FOREST

Whilst streamflow totals are observed to eventually return to preclearing levels where regrowth is allowed, the conversion of native forest to other types of vegetation cover may produce permanent changes. For example, permanent increases in annual water yield are associated with the conversion of deciduous or evergreen native forest to agricultural cropping. Reported increases range from 60-125 mm year^{-1} under humid warm temperate conditions to 450mmyear–^{1} in the equatorial tropics. The diminished water use of annual crops compared with that of a full-grown forest reflects the diminished capacity of low vegetation not only to intercept rainfall but also to extract water from deeper soil layers during periods of drought.

The former relates primarily to the lesser aerodynamic roughness of short annual crops (and possibly to their smaller leaf area and thus storage capacity as well), whereas the reduced water uptake of crops reflects their more limited rooting depth. For the same reasons, conversion to pasture generally produces permanent increases in streamflow as well, although the magnitude of the increase depends on precipitation patterns and elevation. Interestingly, Hibbert (1969) noted that replacing deciduous hardwood forest in the south-eastern USA by vigorously growing *Festuca* grass did not lead to increased water yields, although increases of up to 125 mm year^{-1} were observed after the productivity of the grass declined.

Such results can be explained in part by the fact that serious water shortages that could have reduced water uptake by shallow-rooted plants do not develop in the rainy climate at Coweeta, whereas in addition the contrast between forest and grass was lessened further by the deciduous character of the forest. In contrast, the conversion of mixed deciduous hardwoods by eastern white pine *(Pinus strobus)* in the same area brought about a reduction of 250mmyear$-^{1}$ after 25 years.

The decline in flow was ascribed both to a steady increase in intercepted rainfall as the pines grew older and to higher transpiration by the evergreen pine trees during the dormant season.

Similarly, the higher total water use *(ET)* of mature *Pinus radiata* plantations in south-eastern Australia compared with that of native eucalypt forest has been explained in terms of the distinctly higher rainfall interception by the pines. However, where pine plantations replaced non-deciduous native forests elsewhere in the world, streamflows have been reported to return to preconversion levels within 8 years, such as in Kenya and New Zealand.

The results presented thus far pertain mostly to small headwater catchment areas involving a unilateral change in cover. Although these experiments provide a clear and consistent picture of increased water yield after replacing tall vegetation by a shorter one, and *vice versa,* such effects are more difficult to discern in large river basins having a variety of land-use types and temporal changes therein.

In addition, there are the complications associated with strong spatial and temporal variability in rainfall, especially in tropical areas with largely convective rainfall, and the often large-scale withdrawals of water for municipal, agricultural and industrial purposes in populated areas. Qian (1983) was unable to detect any systematic changes in streamflow from catchments ranging in size from 7 to 727km^2 on the island of Hainan, southern China despite a 30 per cent reduction in tall forest cover over three decades. Dyhr-Nielsen (1986) arrived at the same conclusion for the 14 500 km^2 Pasak River basin in northern Thailand, which lost 50 per cent of its tall forest cover between 1955 and 1980.

On the other hand, Madduma Bandara and Kurupuarachchi (1988) observed an increase in averaged annual flow totals for the 1100 km^2 upper Mahaweli catchment in Sri Lanka over the period 1940-80, despite a weak negative trend in rainfall over the same period. Although both trends were not statistically significant at the 95 per cent level, the associated increase in annual run-off ratios was highly significant.

The increased hydrological response was ascribed to the widespread conversion of tea plantations (not forest) to annual cropping and home gardens without appropriate soil conservation measures (Madduma Bandara and Kurupuarachchi 1988).

However, when analysing a longer time series (1940-97) for the 380 km^2 upper Nilwala catchment in Sri Lanka, which experienced a 35 per cent reduction in forest cover during the observation period, a less consistent picture was obtained. Such contrasting findings illustrate the need for high-quality rainfall and streamflow data in historical hydrological data analyses, not only in the context of evaluating the effects of land-use transformations but also those associated with climatic change.

EFFECTS OF FOREST CLEARING ON STREAMFLOW REGIMES

In areas with seasonal rainfall, the distribution of streamflow throughout the year is often of greater importance than the total annual amount *per se*. Reports of greatly diminished streamflows during the dry season after forest clearance abound in the literature, particularly in the tropics. At first sight, this seems to contradict the evidence presented earlier, that forest removal leads to higher water yields, even more so because the bulk of the increase in flow after *experimental* clearing is observed during baseflow or dry-season conditions. However, the controlled conditions imposed during catchment experiments differ from those encountered in many real-world situations.

GREEN SPACE DEVELOPMENT AND MANAGEMENT

To develop, implement and coordinate the urban forestry programmes for Metro Manila, DENR created the Urban Forestry Division (UFD) in 1988, now Urban "Forestry and Law Enforcement Division (UFLED), under the Forest Management Services (FMS) of DENR – NCR. The Urban Forestry Section under the UFLED has 2 units: the Planting Stock Unit, which is responsible in planting stock production and distribution, and the Cooperative Planting Unit which is tasked to monitor collaborative UF programmes like LKP, CGP and Adopt – A – Street/Park Programmes. The Urban Forestry Section is also currently implementing the OPLAN SAGIP PUNO programme. The urban forestry/greening programmes of DENR- NCR was conceived and adopted to make Metro Manila into a green metropolis. Among its activities include production of planting stocks; establishment and maintenance of mini – forests; greening of main thoroughfares, side streets and islands; establishment of joint nurseries with LGUs, NGOs, POs, schools; providing technical assistance and training on proper site preparation, choice of species and proper planting and maintenance; harnessing the cooperation and involvement of the public via information campaign; periodic assessment of greening activities; processing and issuance of balling permits, and implementation of Oplan Sagip Puno programme.

At the city or municipal government level, specific urban forestry/greening offices were created under the Mayor's office. Some of these offices are ad – hoc in nature while the others are permanently institutionalized in the city government structure. The greening offices of all the 6 cities have a mandate on green space development and management although in 3 cities (Makati,

Pasig, Mandaluyong), they also included cleaning, waste management and pollution monitoring and control as their other mandates. In addition to the greening office, each city also has either a committee or Tasks Force created to enhance active participation of other sectors in the greening and cleaning activities.

The capabilities of these greening offices in terms of manpower, available facilities and financial resources were also assessed by Palijon (2000). He found that: a) majority of these offices felt the need for additional technical staff to fully implement their greening programmes such as horticulturist, forester/ arborist, and landscape architect; b) the greening offices have inadequate facilities, equipment and tools needed for their programmes; and c) all cities, except Makati, have insufficient and unsustainable budget for their greening programmes. In the case of the other mega-cities, the DENR regional office does not have specific urban forestry unit unlike in the NCR. Their urban forestry programme activities are being handled by the Reforestation Section under the Forest Resources Development Division of the Forest Management Services, except in Cagayan de Oro where a focal Programme Unit for Urban Forestry was recently created by the RED attaching such unit directly to the RTD-FMS. At the city government level, the City ENRO, Clean and Green Office, Task Force Clean and Green and Cleaning and Beautification Committee are usually the offices or bodies attached or created under the Mayor's Office responsible for management of urban forestry programmes just like those of the NCR.

PHYSICAL ACCOMPLISHMENTS OF URBAN FORESTRY PROGRAMME

The total number of seedlings planted in the urban greening programme by city/municipality in Metro Manila from 1988 to 2002. The 17 cities and municipalities comprising Metro Manila (NCR) have an aggregate total of 2,212,488 seedlings planted. The highest number of seedlings planted was in Quezon City (0.72 million seedlings or 34 per cent of total), followed by Manila City (0.31 million seedlings or 14 per cent of total). On the other hand, the total population of Metro Manila in 2000 was 9.93 M people.

Based on the statistics, the estimated tree to person ratio is 1:6 or 1:9 assuming 80 per cent or 50 per cent survival of the seedlings planted, respectively. These ratios are short of the 1:4 target ratio set by MPFD. However, the estimated number of trees is most likely an underestimation because the trees already existing prior to 1988 were not accounted or tallied. For the other mega-cities, the attainment of 1:4 ratio can not be ascertained due to absence of data on the number of seedlings planted on different years and the absence of inventory data on the number of trees already existing in these areas.

Accomplishment for Mini Forests and Parks

The minimum number of mini-forests or parks targeted by the MPFD for Metro Manila was 60 for the 8.2 million residents and this was based on the assumption that there should be 1 mini-forest or park per 100,000 – 150,000 residents. Since the present population of Metro Manila (NCR) is about 10M, there should be at least 67 mini-forests or parks already established and maintained.

As of 1994, there were already 472 parks established in different parts of Metro Manila, which is more than enough (*i.e.* 7 times more) compared to the target ideal number. Almost 50 per cent of the total number of parks in Metro Manila were established in Quezon City. Between 1996 to 2001, eight more mini-forests were established in Metro Manila covering about 18.20 hectares. Hence, for the entire Metro Manila, a total of 480 mini-forests or parks have already been established as of 2001.

Compound Planting

As envisaged in the MPFD, at least 570 compounds or grounds should have been planted or landscaped by year 2000 in all six priority cities. Since there is no breakdown per city, the target maybe equally allocated at 95 compounds or grounds per city. There is no available information on the number of compounds or grounds planted or landscaped in the different mega-cities. Nevertheless, for Metro Manila, data is available on the number of seedlings planted. For the past 14 years, about 1.1 million seedlings have been planted on different types of compounds mostly in the government offices and schools in Metro Manila.

Greenbelt Development

For Metro Manila, the MPFD targeted 100 km of roadsides planted or greenbelt developed from 1990 – 2000. Since there is also no data available on actual length of roads planted or greenbelt developed, we can use the available data on number of seedlings planted along thoroughfares and streets for extrapolation. There were 673,813 seedlings planted along major thoroughfares and barangay roads of Metro Manila for the past 14 years. Due to limited space along road shoulders, it is assumed that there will be a single line planting on each side of the road. At 2m spacing, there will be 1,000 seedlings required per km of road. Assuming these 673,813 seedlings planted included 50 per cent replanting, it is estimated that about 337 km of roads and thoroughfares should have been planted in Metro Manila. This is 3.4 times more than the target of MPFD.

For Iloilo City, the DENR-FMS Region 6 reported 34 adoptors (government agencies, universities, private companies, NGOs, etc.) involved in Adopt-A-Street/Park Programme since 1994. Of the 34 adoptors, 31 adopted streets spanning 17 km while 3 adopted perimeters of parks covering about 1 km. Since the MPFD's roadside planting target from 1996 to 2000 is 35 km, only about 50 per cent of the target was accomplished.

In the case of Cagayan de Oro City, there were 10.5 km roadside planting done in 1995 to 1999 in barangays Balubal and Indahag and along Lumbia Airport Road (DENR-FMS Region 10). On the other hand, the City ENRO conducted 8 km highway tree planting/urban greening in barangays Cugma, Tablon, Bugo and other urban barangays (City ENRO – Cagayan de Oro City). Thus, the DENR and City ENRO's roadside planting efforts fall short (*i.e.* only 46 per cent accomplishment) of the MPFD's 1996-2000 target of 40 km. However, for the year 2001-2005, the MPFD's target of 20 km greenbelt development was already accomplished through the tree and bamboo planting project undertaken in 2002 along 20 km riverbanks in barangays Balulang, Lumbia, Bayanga, Mambuaya and Dansolihon through a contract awarded by the ENRO to MARBEMCO, a people's organization (City ENRO - Cagayan de Oro, 2002).

URBAN FORESTRY DEVELOPMENT AND MANAGEMENT

The urban forest is all of the woody vegetation growing in an urban area, including trees, shrubs, and vines found along city streets, public parks and private property. The City is responsible for managing an urban forest that contains over 40,000 street trees and 6,600 park trees (mowed areas).

There are several policies, programmes and projects issued and implemented for the past several decades which is an indication of continuing concern on the deterioration of urban environment. It is noticeable that new policies and programmes evolve whenever there is a change in administration (a common phenomenon in the Philippines) indicative of lack of continuity of previous initiatives. The major policies and programmes related to urban forestry are chronologically listed below:

- PD 1153 of Pres. Marcos dated 1976 (Tree Planting Decree to support PROFEM)
  - Requires all able-bodied Filipinos 10 years old and above to plant a tree per month for 5 consecutive years.
  - Certificates of planting and survival— requirement for graduation from school, renewal of job appointment and business permit and approval of retirement from service.
  - "Halamanan ng Bayan" launched by MHS to support this programme. It required each city or municipality to put up a nursery, garden and park.
  - Repealed by EO 287 dated July 25, 1987 because of dictatorial provisions and harsh penalty.
- PD 953 of Pres. Marcos dated July 6, 1976 (Greening of Private Lands Including Residential Subdivisions)

  - Requires private landowners to plant trees extending at least 5 m on each side of the rivers/ creeks.
  - Developers or owners of residential subdivisions and commercial/ industrial lots to set aside 30 per cent of total area as open spaces for parks and recreational areas.
  - Penalizes unauthorized cutting, destruction or injury inflicted on naturally-growing or planted trees or vegetations in any public places.
- LOI 1312 of Pres. Marcos dated April 23,1983 (Establishment and Development of Local Government Forest or Tree Parks Throughout the Philippines).
  - Requires each barangay, municipality or city to establish and maintain at least one forest or tree park of considerable size.
  - MNR (now DENR) to allocate public lands for this purpose and to provide technical assistance and seedlings needed.
  - MHS to ensure that establishment of forest or tree parks is included in the land use plan of each barangay, municipality or city.
  - MILG (now DILG) to appropriate funds and implement establishment and maintenance activities.
- Memo Order Nos. 198 and 199 of Pres. Aquino dated November 9, 1988 (Luntiang Kamaynilaan Programme (LKP)/ Hardin ng Bayan Programme).
  - Issued to help insure healthy environment in Metro Manila (MM) and to serve as model programme for other cities/municipalities.
  - Anchored on the "Hardin ng Bayan" concept wherein each city or municipality should have gardens or parks of their own, transforming MM into a garden metropolis with lush vegetations, cool and fresh air like the countryside.
  - Objective- to plant 2 million trees in 2-3 years and achieve a desired 1:4 tree-man ratio.
  - For efficient, effective coordinated implementation, an Inter-Agency Committee (IAC) was formed: Co-chair- DENR and MMA (now MMDA); members- DPWH, DOTC, Metro Police Force, DOT, OPS and PMS.
- Memo Cir. No. 5 of Pres. Ramos dated August 27, 1992 (Clean and Green Programme).
  - Similar to LKP (same IAC composition except MMDA as chair/ lead agency) but wider in scope (not only greening but also cleaning activities)
  - Objective- massive planting (0.5 million trees/year or 2.5 million trees in 5 years from 1993-1997) to achieve the ratio of one tree for every 4 persons.

  - Although focused in MM, CGP has nationwide coverage and encouraging cities and municipalities to join nationwide contest for cleanest and greenest city or town.
- EO No. 113 of Pres. Ramos dated July 22, 1993 (Multi-sectoral Tree Planting Activities in Support of ENR Programmes/ECOREV).
  - Scope/ Objective – regreening and rehabilitation of all open and denuded lands of public domain, idle lands, private lands and other suitable areas (both urban and rural) including rehabilitations of coastal and marine areas.
  - DENR to identify, assess and designate suitable area for planting and management and to provide technical assistance to participating agencies.
  - LGUs implement the programme in their respective level and set up counterpart funds.
  - Private sector participation encouraged via MOA or other appropriate arrangements with DENR.
- EO No. 118 of Pres. Ramos dated August 12, 1993 (Mandating the active participation of all government agencies nationwide in urban greening through an Adopt-A-Street/Park Programme)
  - Objective – greening of streets and parks in urban centers.
  - Requires all government offices and government owned/controlled corporations to adopt a street or park in coordination with concerned LGUs, NGOs and private sector by planting appropriate species and maintaining them for at least 5 years using their own funds/resources. and other resources.
  - DENR to manage and coordinate the programme through a designated National Coordinator.
  - Project to be turned over to concerned LGU for maintenance and protection.
- DENR-DILG-DPWH-CSC Joint Memorandum Circular No. 1 dated December 17,1993 (Implementing Guidelines for EO 118-Adopt-A-Stree/Park Programme)
  - Described the roles of each participating agency and outlined the schemes in the identification, selection and adoption of a street or park to be developed.
  - DENR to provide assistance to "adopters" in selecting suitable streets or park sites, in providing necessary planting materials and in monitoring performance.
- OPLAN SAGIP PUNO Programme of FMS-NCR/DENR launched on June 5, 2000.
  - Conceived as a component of "Lets Go Green Programme" of former DENR Secretary Antonio Cerilles.

- Application of appropriate silvicultural treatments (*e.g.* removal of nails, wires/cables, water sprouts; surgical treatment of injured stem or root) to prolong life span and promote good health and vigour of trees planted in parks and along thoroughfares and streets in MM.
- Supplemented by public awareness campaign.
- DENR enters into MOA with participating agencies (*e.g.* subdivision homeowners association, city/ municipal government, NGOs, etc.)
- DENR's role — conduct inventory and assessment of damaged/ injured trees; undertake appropriate silvicultural treatments; conduct information dissemination and training on tree care and maintenance; provide technical assistance and planting materials to sustain the project.
- LGU's role — provide tree care and maintenance crews to sustain the project; assist DENR in information dissemination on maintenance and protection of trees.

- Proclamation No. 396 of Pres. Arroyo dated June 2, 2003 (Enjoining the active participation of all government agencies including government-owned or controlled corporations, private sector, schools, civil society and citizenry in tree planting activity and declaring June 25,2003 as Philippine Arbor Day).
  - Objectives- to promote multi-sectoral participation in tree planting nationwide; to develop greater awareness on the importance of trees in environment, health and human life.
  - Participating agencies, LGUs, schools, etc. to identify areas to be planted in coordination with agencies which have jurisdiction over such areas *e.g.* DENR in case of public lands, LGUs in areas within their jurisdiction, DND for military lands reservation, DOT for ecotourism areas, etc.
  - DENR, LGUs and schools — to establish and maintain nurseries.
  - Respective participating agency/ instrumentality — to maintain and protect the planted seedlings.
  - DENR — to provide technical assistance to all participants.

In general, the following goals and objectives are common to the Urban Forestry (UF) policies and programmes described above:

- to provide/maintain green, clean and beautiful environment;
- to promote public awareness on the importance of trees (promote environmental consciousness);
- enhance people's participation in the programme;
- promote multi-sectoral collaboration, cooperation and support; and
- in the case of LKP and CGP, the specific objective is to attain a 1:4 tree to person ratio to sustain ecological balance.

As strategy to enhance successful implementation of the project, DENR is usually tasked to provide technical assistance in planting, site and species selection, and maintenance operations, including provision of the planting stocks. Understandably, the DENR is also looked up to as the lead agency when inter-agency collaboration is involved in the programme. On the other hand, the city, municipal and barangay governments, which have jurisdiction over the project site, are usually tasked to maintain and protect the tree parks established and streets planted. They are also required to provide counterpart funds and other resources needed for these projects.

At the end of each programme, there seems to be no serious post – project accounting or evaluation of outputs and accomplishments, including evaluation of success and failures. This may be attributed to the fast rate of turn – over of urban forestry/greening programmes being implemented. Another reason maybe lack of manpower and resources to monitor all the projects. For instance, in the case of Metro Manila, the Urban Forestry and Law Enforcement Division Office of FMS – NCR/DENR only has a small unit (Cooperative Planting Unit) under the Urban Forestry Section which is tasked to do the monitoring activities. Needless to say, the synthesis of lessons learned is an important input for planning and formulation of new programmes (*i.e.* we do not have to " reinvent the wheel" so to speak).

DECENTRALIZATION OF FOREST MANAGEMENT

Although a majority of forests continue to be owned formally by government, the effectiveness of forest governance is increasingly independent of formal ownership. Since neo-liberal ideology in the 1980s and the emanation of the climate change challenges, evidence that the state is failing to effectively manage environmental resources has emerged. Under neo-liberal regimes in the developing countries, the role of the state has diminished and the market forces have increasingly taken over the dominant socio-economic role. Though the critiques of neo-liberal policies have maintained that market forces are not only inappropriate for sustaining the environment, but are in fact a major cause of environmental destruction. Hardin's tragedy of the common (1968) has shown that the people cannot be left to do as they wish with land or environmental resources. Thus, decentralization of management offers an alternative solution to forest governance.

The shifting of natural resource management responsibilities from central to state and local governments, where this is occurring, is usually a part of broader decentralization process. According to Rondinelli and cheema (1983), there are four distinct decentralization options: these are: (i) Privatization – the transfer of authority from the central government to non-governmental sectors otherwise known as market-based service provision, (ii) Delegation – centrally nominated local authority, (iii) Devolution – transfer of power to locally acceptable authority and (iv) Deconcentration – the redistribution of authority from the central government to field delegations of the central government. The major key to effective decentralization is increased broad-based participation in local-public decision making. In 2000, the World Bank report reveals that local government knows the needs and desires of their constituents better than the national government, while at the same time, it is easier to hold local leaders accountable. From the study of West African tropical forest, it is argued that the downwardly accountable and/or representative authorities with meaningful discretional powers are the basic institutional element of decentralization that should lead to efficiency, development and equity. This collaborates with the World Bank report in 2000 which says that decentralization should improve resource allocation, efficiency, accountability and equity "by linking the cost and benefit of local services more closely".

Many reasons point to the advocacy of decentralization of forest. (i) Integrated rural development projects often fail because they are top-down project that did not take local people's needs and desire into account. (ii) National government sometimes have legal authority over vast forest area that they cannot control, thus, many protected area project result in increased biodiversity loss and greater social conflict.

Within the sphere of forest management, as state earlier, the most effective option of decentralization is "devolution"-the transfer of power to locally accountable authority. However, apprehension about local governments is not unfounded. They are often short of resources, may be staffed by people with low education and are sometimes captured by local elites who promote clientelist relation rather than democratic participation. Enters and Anderson (1999) point that the result of community-based projects intended to reverse the problems of past central approaches to conservation and development have also been discouraging.

Broadly speaking, the goal of forest conservation has historically not been met when, in contrast with land use changes; driven by demand for food, fuel and profit. It is necessary to recognized and advocate for better forest governance more strongly given the importance of forest in meeting basic human needs in the future and maintaining ecosystem and biodiversity as well as addressing climate change mitigation and adaptation goal. Such advocacy must be coupled with financial incentives for government of developing countries and greater governance role for local government, civil society, private sector and NGOs on behalf of the "communities".

6

Principles of Forest Pest Management

Interest in protecting forests from insect, disease, weed and vertebrate pests has increased in recent years.

This has come about largely because of:

- Increased awareness of the destructive capacities of pests;
- The heavy toll they take on supplies of commercial and recreational timber;
- Environmental concerns;
- Effects on threatened and endangered species; and
- Availability of new, specific pesticides.

Forest managers have come to realize that much of the damage caused by pests could have been avoided. With adequate knowledge of pest identification and biology, combined with good forestry management practices, it may be possible to prevent or at least reduce losses due to pests. Trees in a vigorous condition are much better able to withstand damage by pests than trees already under stress.

We have learned that using a combination of prevention and control methods is the best approach to pest problems. The planned strategy of combining the best methods is called Integrated Pest Management (IPM) and is discussed in the "Applying Pesticides Correctly" core manual. Pest management should be a part of an overall forest management plan. The need for pest control treatments can often be minimized through wise, long-term forestry practices. The pest control method(s) chosen will depend upon the kind and amount of control necessary, balanced with costs and benefits within legal, environmental and other constraints. The most important principle of pest control is to use a control method only when necessary to prevent unacceptable levels of damage. Even though a pest is present, it may not be necessary to control it. It may cost more to control the pest than to cover damage or losses.

Before making management decisions, managers should evaluate potential pest impacts within the context of the ecosystem in which the organism occurs, as well as the population dynamics of the organism. Will the impact of an organism increase, decrease or maintain its level of damage over time? What part(s) of the tree does the pest affect? How many trees are or will potentially be affected? What will be the longterm impact of these organisms? Does the organism cause permanent or only temporary damage? Insects such as the southern pine beetle damage the cambium layer and introduce fungi that almost always cause tree death. In contrast, many foliage feeding insects cause one-time defoliation from which the tree can recover. Most trees can withstand complete one-time defoliation without significant longterm impact on tree health. However, an organism that has the potential to cause multiple defoliations (such as the gypsy moth) can have a much more detrimental impact on tree and forest health.

BEFORE CHOOSING A CONTROL METHOD(S)

- Correctly identify the organism to ensure it is a pest.
- Monitor the pest populations and determine the likelihood of economic damage.
- Review available control methods.
- Know and follow local, state and federal regulations that apply.
- Evaluate the benefits and risks of each available treatment method or combination of methods.
- Determine whether there are any threatened or endangered species in the area to be treated.
- Choose the method(s) that are effective yet will cause the least harm to you, others and the environment.
- Correctly carry out the control practice(s) and keep accurate records.

If other management options do not yield satisfactory results, you may need to apply a pesticide to control an undesirable organism (pest) in the environment. The challenge is to use pesticides in a manner that will cause the least harm to non-target organisms in forests, seed orchards and nurseries, while achieving the desired management goal.

Pesticides are tested and labelled for specific pests, crops and for land-use situations. Use of insecticides, fungicides and herbicides is common in managed seed orchards, forest nurseries, intensive short-rotation plantations, and in Christmas tree production. In general, the most commonly used forest pesticides are herbicides used for site preparation, herbaceous weed control, and in pine release treatments. Insecticides are seldom used in general forest management because of high treatment costs and because some pest insects are highly mobile.

Currently, the only disease control treatment common in general forestry field applications is for annosus root rot. Vertebrate animals are sometimes controlled through trapping or hunting, but repellents and poison baits may be employed.

FOREST PEST CONTROL

INSECTS

Insects are the most destructive agents affecting forest and shade trees in the South. Tree roots, stems, limbs, needles, leaves of healthy or weakened trees, or logs waiting to be sawed into lumber are all subject to attack. Insects (Class: Insecta) are by far the most numerous animal life inhabiting the forest. They have become well adapted to their surroundings and occupy a wide variety of ecological niches. Although the majority of insect species are either beneficial or innocuous, some are exceedingly harmful. Insect outbreaks that cause economic damage to forests vary greatly in frequency, size and duration. Fortunately, most outbreaks are small and short-lived, and usually consist of one or a few spots in a stand or region. Others, however, may expand and encompass hundreds or thousands of acres and can last for several years.

Managers can reduce the risks and incidence of insect attack by maintaining healthy and vigorously growing stands and trees. Research is being conducted to determine what conditions are conducive to forest insect outbreaks. This research may lead to improved control measures.

LOSSES CAUSED BY FOREST INSECTS

These data are summarized from Special Bulletins published annually by The University of Georgia, College of Agricultural and Environmental Sciences, Department of Entomology, Insect Survey and Losses Committee. These figures have not been adjusted for inflation.

Table. Estimated Average Yearly Losses and Control Costs of Forest Insects in Georgia

Rank	Insect	Cost of Control	Damage Loss	Total Cost
1.	Southern pine beetle	$317,600	$4,680,400	$4,998,000
2.	Pine tip moths[1]	$916,000	$2,310,000	$3,226,000
3.	Defect & degrade[2] causing insects	$90,000	$2,874,000	$2,964,000
4.	Seed & cone[3] insects	$80,400	$2,514,200	$2,594,600
5.	Ips spp. beetles[4] and black turpentine beetle	$485,000	$2,016,400	$2,501,400
6.	Reproduction weevils[5]	$1,153,000	$928,000	$2,081,000
7.	Other insects[6]	$83,800	$1,215,000	$1,298,800
8.	Gypsy moth	$94,000	$--	$75,200
	Totals	$3,201,000	$16,538,000	$19,739,000

Notes:

1. Includes Nantucket pine tip moth and pitch pine tip moth.
2. Includes carpenter ants, ambrosia beetles, lepidopterous oak borers, shothole borers and various other cerambycid, buprestid and scolytid beetles.

3. Includes coneworms, seedworms, seed bugs and cone beetles.
4. Ips avulsus, I. grandicollis, I. calligraphus and I. pini.
5. Pales weevil and pitch-eating weevil.
6. Primarily aphids, scale insects, sawflies and lepidopterous hardwood defoliators including eastern tent caterpillar, forest tent caterpillar, fall webworm, oak skeletonizer and various Anisota spp.

BARK BEETLES

Bark beetle populations vary tremendously between years and between locations. The important bark beetles in the South attack pines and belong to the Family Scolytidae. However, some species such as the native elm bark beetle Hylurgopinus rufipes, the small European elm bark beetle Scolytus multistriatus (Marsham), and the hickory bark beetle S. quadrispinosus Say are hardwood pests. The remainder of this bark beetle discussion will be about southern pine beetles, black turpentine beetles and Ips engraver beetles that attack southern pines.

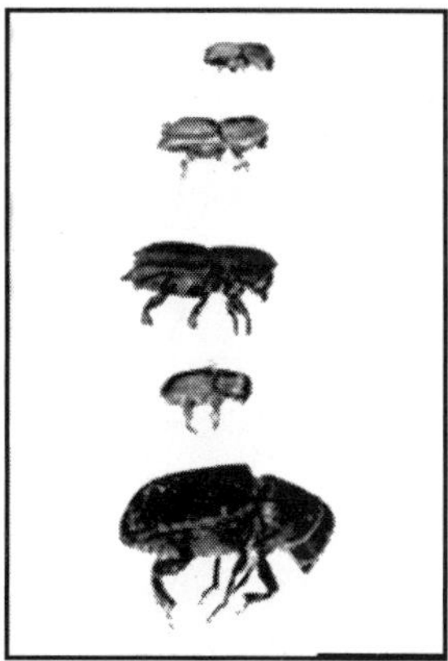

Fig. The Southern Pine Bark Beetles. Top to Bottom: *Ips Avulsus, Ips Grandicollis, Ips Calligraphus, Dendroctonus Frontalis* (SPB), *Dendroctonus Terebrans* (BTB), Gerald J. Lenhard, LSU.

Large numbers of attacking bark beetle adults can often overwhelm a tree's natural defences, lay eggs and successfully initiate an infestation. After the eggs hatch, the grub-shaped larvae can girdle the tree by feeding under the bark. Blue-stain fungi are also carried on the bodies of most species and introduced into the tree during adult attack. Proliferation of these introduced fungi in the water-conducting tissues hastens the death of the infested trees.

Pines under stress are particularly susceptible to bark beetle attacks, especially if beetle outbreak conditions exist. Bark beetle attacks can be recognized by boring dust and pitch tubes on the outside of the bark, characteristic galleries under the bark, and adult beetles and larvae in the inner bark. After a tree is successfully attacked, the foliage fades from bright-green to yellowish-green to red. These "faders" are generally the first apparent sign of a bark beetle attack. Unfortunately, foliage colour change often does not occur until well after the tree is dead and the beetles have completed their development and have left the tree.

Bark beetle brood development time ranges from 25 to 120 days, depending upon species and temperature. It is common for more than one species of bark beetles to infest individual trees. Since many other insects are associated with dead and/or dying trees, make positive identification of the insects before you take any remedial actions.

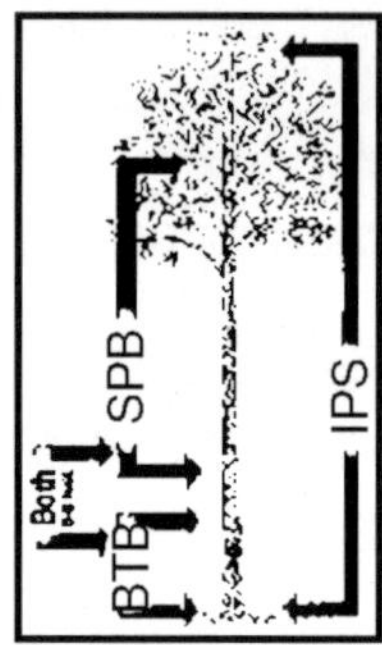

Fig. Sections of the Trunk which Three Types of Bark Teetles Attack.

Reduce the potential for bark beetle attack by ensuring that trees are rapidly growing and healthy. Removing or treating lightning and storm damaged trees promptly and maintaining proper stand densities can reduce the likelihood of bark beetle attack. For specific management and control practices, contact your county Extension Service agent or State Forestry office.

SOUTHERN PINE BEETLE

The southern pine beetle (SPB), Dendroctonus frontalis Zimmermann, is the most destructive of the eastern species of pine bark beetles. It is a small reddish-brown to black beetle, about 1/8 of an inch long. The rear end of the body is rounded. SPB normally infest boles of trees from the base to the crown, with initial attacks at mid-bole or higher. The life cycle generally requires 35-60 days to complete. There may be up to six generations per year. Small pitch tubes, usually less than 2 inch in diameter, are often present at the site of adult attacks.

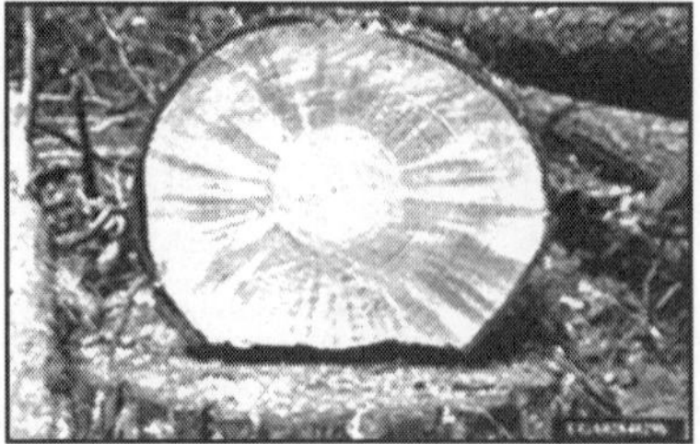

Fig. *Southern Pine Beetle*: Pine Killed by Beetle with Blues Stain Fungus, A Cross Section of Log, Ron Billings, Texas Forest Service.

Adult SPB bore directly through the bark and mate. The females excavate the characteristic S-shaped, crisscrossing egg galleries in the inner bark. Eggs, which are deposited in niches on either side of these galleries, hatch into small, legless grubs within 4-9 days. The grubs mine for a short distance before boring into the outer bark where they pupate. Galleries are usually filled with larval fecal material and boring dust. Soon after a tree is attacked, all the needles turn yellow and then brown. Drought seems to trigger major outbreaks of this insect. In addition to direct damage caused by the larval feeding, SPB introduce stain-causing fungi, which invade the tree and hasten death. Once southern pine beetles have successfully infested a tree, no remedial treatments are available to prevent the death of the tree.

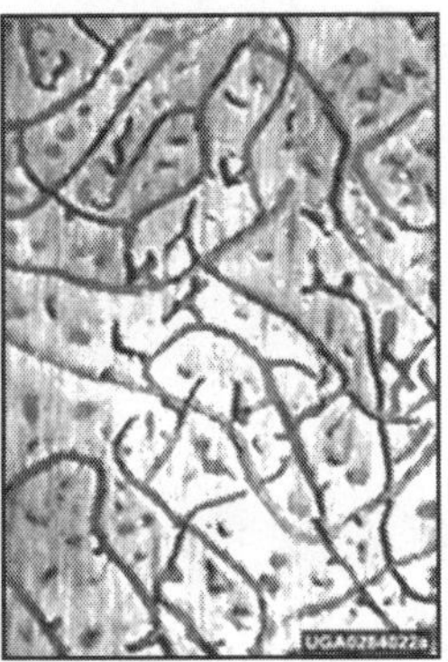

Fig. *Southern Pine Beetle*: 'S' Shaped Galleries Under Bark, Ron Billings, Texas Forest Service.

BLACK TURPENTINE BEETLE

The black turpentine beetle (BTB), Dendroctonus terebrans (Oliver) is the largest of the major southern bark beetles. It is ¼ of an inch or more in length, with a rounded rear end, and is reddish-brown to black in colour. Large purplish coloured pitch tubes (50 cent piece in size) are often present at the site of adult attacks.

Fig. Southern Pine Beetle Pitch Tubes on Loblolly Pine, Jim Meeker, Florida Dept. of Agriculture and Consumer Services.

BTBs attack fresh stumps and living trees by boring through the bark and constructing galleries on the face of the sapwood where 50-200 eggs are laid in a group. BTBs attacks usually only occur on the trunks of trees up to a height of eight feet. After hatching, the white larvae feed on the inner bark. Unlike southern pine beetles and Ips engraver beetles, black turpentine beetles do not introduce blue stain fungi into the tree and developing larvae feed in "patches" rather than completely encircling the tree. However, when several broods occur at about the same height, the feeding larvae may completely girdle and kill the tree. Without prompt treatment, from 70 to 90 per cent of the trees attacked by the BTB die. The BTB life cycle takes from 2 ½ to 4 months, depending on the temperature. In the South, there are usually two generations and part of a third each year.

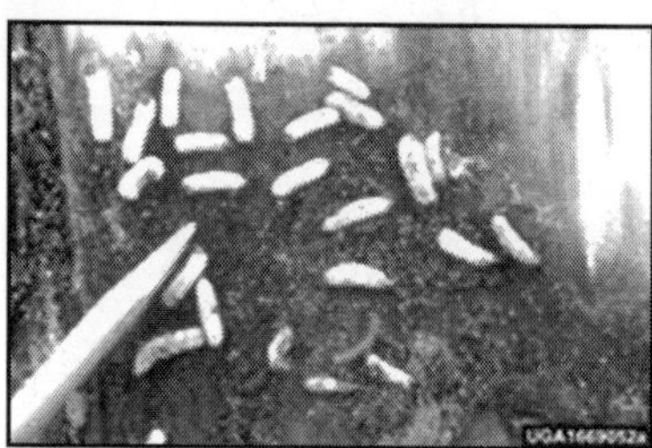

Fig. Black Turpentine Beetle Larval Feeding Patch, NC State University Archives.

Fig. Black Turpentine Beetle Pitch Tube, NC State University Archives.

IPS BEETLES

The four species of Ips beetles commonly found in the South (Ips grandicollis, I. calligraphus, I. avulus, and I. pini) vary from 1/10 to ¼ of an inch long, and are yellowish, dark reddish-brown to black (Figures). They are easily recognized by their scooped out posteriors which are surrounded by varying numbers of tooth-like projections.

In hot weather, it may take as few as 25 days to complete one generation. Populations of these beetles increase rapidly during favourable conditions. Ips spots usually contain only one to a few trees killed, but under favourable conditions Ips can become epidemic and kill many trees. The fully grown, grublike larvae are yellowish-white and vary from ¼ to 1/3 inch long. Gallery patterns are more or less Y- or H-shaped except for I. avulus, which deviates from these patterns. Small, white eggs are laid singly in small egg niches cut along the main tunnels. Larval feeding tunnels are usually filled with boring dust and frass (excrement).

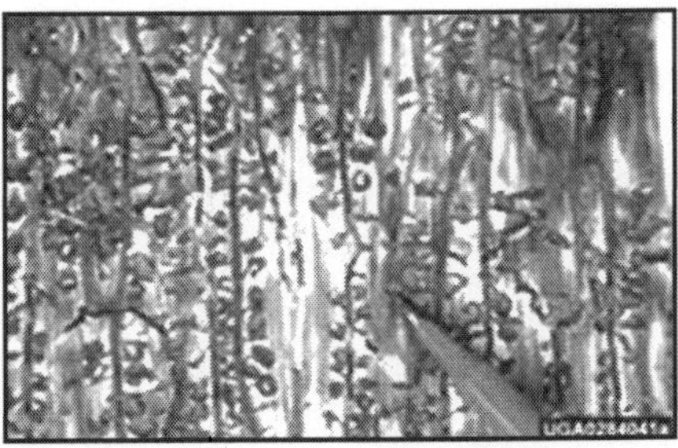

Fig. *Ips Avulsus* Adult and Larval Galleries, Ron Billings, Texas Forest Service.

Adults girdle trees quickly as they construct their egg galleries in the inner bark. The tree's death is usually hastened by the introduction of blue-stain fungi which blocks the flow of sap. Small reddish pitch tubes are frequently the first sign of an attack. These tubes are usually absent in trees suffering from drought. As with SPB, once Ips beetles have successfully infested a tree, the tree cannot be saved!

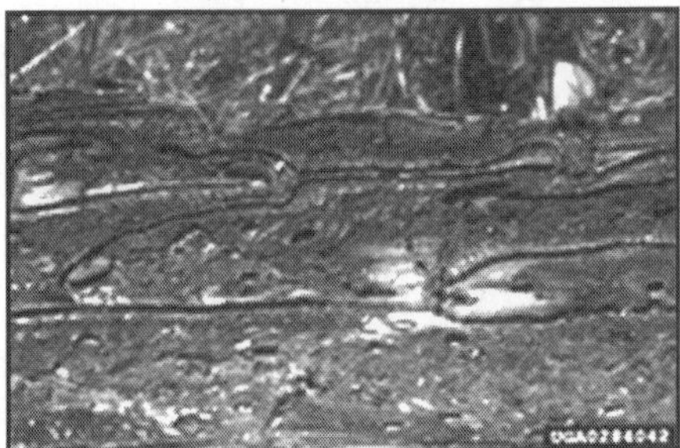

Fig. Eastern Fivespined *Ips* Adult and Larval Galleries, Gerald J. Lenhard, LSU.

PINE BARK BEETLE CONTROL

Several methods reduce the likelihood of expanded attacks by bark beetles. Most of these are based on good silvicultural practices to keep trees rapidly growing or by reducing stress on trees.

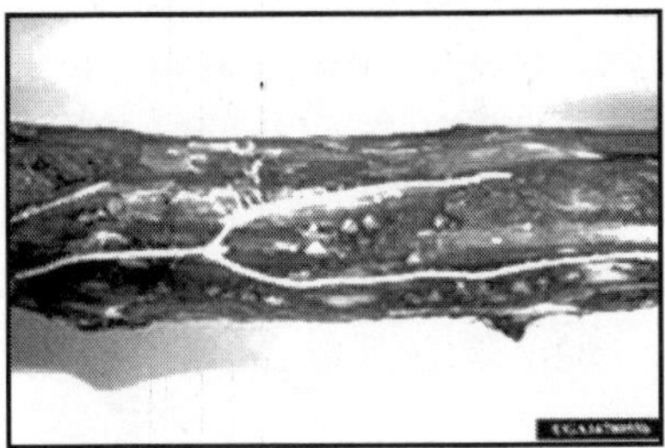

Fig. *Ips Pini* gallery on Small Pine, David McComb, USDA Forest Service.

Techniques currently used are:

- Prompt removal (and/or treatment) of trees damaged by lightning, storms or by construction.
- Salvage removal.
- Cut and leave.
- Pile and burn.
- Insecticide treatments.
- Cut and spray.
- Behavioural modifying chemicals.

Prompt Removal of any damaged trees, whenever possible, significantly reduces the likelihood of successful bark beetle attacks. Since bark beetles are attracted to odours exuded from damaged trees, initial attacks in an area often occur on damaged trees. As the beetle brood matures and exits the infested tree, the infestation frequently expands to other trees in the area.

Salvage Removal is only feasible when a relatively large volume of wood is available and makes the operation cost effective for the logger. Prompt removal of dead and dying trees is essential to prevent significant degradation of the wood.

Steps involved in a successful salvage removal operation include:

- The removal of a 50-100 feet wide buffer strip of green uninfested trees around the most recently attacked trees;
- Removal of recently attacked trees containing developing beetle brood; and
- Removal of older standing trees from which the brood has already emerged.

Cut and Leave is best for controlling small spots of trees (10-50) when salvage is not practical or cost effective. Attacked trees and a border of healthy trees are felled towards the centre of the spot. It is helpful to cut the limbs on the underside of the felled trees so that the trunks are lying on the ground. The increased sunlight and subsequent higher temperatures, along with increased humidity resulting from the trunks lying on the ground, are thought to cause high mortality of the developing brood.

In *Pile and Burn*, trees with live brood are felled, piled and burned. Although effective, this technique requires heavy equipment to pile the trees so that they can be burned. Additionally, managers must be aware of appropriate weather conditions and prescribed fire and smoke management issues.

Insecticides in bark beetle control are more preventative than corrective. Consequently, with the possible exception of the BTB which does not introduce blue-stain fungi into the tree, trees that have been successfully attacked by bark beetles cannot be saved by insecticide applications. However, bark beetles still in or under the bark can be killed by spraying the tree with appropriate insecticides to prevent spread of the attack. Additionally, non-infested high value trees, judged to be at high risk, can be sprayed with an insecticide as a preventative measure against attack. The area of the tree requiring insecticide treatment depends upon the insect species for which the application is being made. The appropriate area of the tree should be thoroughly wetted with the insecticide spray mixture.

Applications made for BTB only require that the lower 8-10 feet of the main bole be treated and can be accomplished with a small hand or backpack sprayer. If applications are being made for SPB or any of the Ips beetles, spray to wet the entire bole of the tree from ground level up to the upper crown, including the base of large scaffold limbs. On larger trees, this requires high-pressure sprayers for thorough coverage.

A *Cut and Spray* method may stop further spread of bark beetle attack. If beetle infested trees can be felled but cannot be hauled from the site or burned, they can be limbed and bucked into workable lengths. Once cut into workable lengths, the tree sections can be turned as they are thoroughly sprayed with an appropriate insecticide.

Behavioural Chemicals are being tested by researchers to develop SPB control tactics by manipulating the various chemicals that the beetles (and their natural enemies) use to orient, attack or disperse their populations. To date, these promising, non-pesticide-based tactics are not yet ready for large scale field implementation.

BORING INSECTS

Included in this group are the insects that infest terminals, shoots, twigs and roots of living trees as well as those that obtain food and shelter from wood. Terminal and shoot insects are of particular importance in the initial stages of forest regeneration and early stand growth. These insects are also of great importance in forest nurseries and ornamental trees. We will discuss in detail the Nantucket pine tip moth, and the white pine and deodar weevils which are frequently encountered in and cause significant damage to pine stands in the South.

Other insects in this category damage or destroy trees that would otherwise produce quality lumber or other wood products. Most insects that cause this damage are borers, either adult or larval stages or both. Most borers are secondary invaders, attacking bark and wood of trees that are seriously weakened, dying, or recently cut. Carpenterworms, ambrosia beetles, oak clearwing borers, metallic wood borers, Columbia timber beetles and southern pine sawyers are some of the pests that cause damage in this category. Trees attacked by these pests are usually scattered so that most control measures are difficult and not economically feasible. Additionally, there are several species of insects that attack, infest and damage wood and wood products that are discussed in the Wood Treatment manual and in other manuals devoted to Pest Control Operator training and are not covered in this manual. Included among the wood product pest group are certain beetles in the family Cerambycidae (notably, the old house borer *Hyloupes bajulus*), ambrosia beetles (Scolytidae and Platypodidae), and the powderpost beetle complex (includes members of the Families Lyctidae, Anobiidae, and Bostrichidae).

NANTUCKET PINE TIP MOTH

The Nantucket pine tip moth, *Rhyacionia frustrana* (Comstock), and its close relative the subtropical pine tip moth, *R. subtropica* Miller are widely distributed in the southern states. The importance of pine tip moths on pine and Christmas tree plantations and nurseries varies widely with tree species, host vigour and environmental factors. Heavily infested trees may be severely stunted or deformed, but mortality is rare. Generally, the tree grows out of the susceptible stage within a few years. All species of pines are attacked except white and longleaf pines, but slash pine is rarely attacked. Loblolly, Virginia and shortleaf pines are most susceptible.

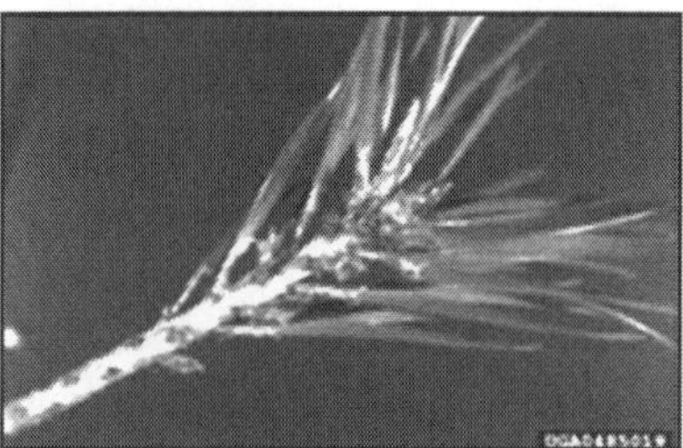

Fig. Nantucket Pine Tip Moth Larva Feeding at Base of Needles, David J. Moorhead, UGA.

The adult moth is mixed gray and shiny copper-coloured, with a wingspan of about ½ inch. The young larvae are light cream-coloured, while mature larvae are light brown and approximately 3/8 inch long. Pupation occurs on the tree in the damaged terminal. Adults begin to emerge on warm days in early spring and begin laying eggs in a few days. Eggs are deposited on needles, stems, developing tips or buds. After hatching, larvae first feed on needle fasicles, then bore into terminals and lateral shoot buds, and finally into stems. The larval period lasts from two to four weeks. There are usually three to four generations per year.

Fig. Nantucket Pine Tip Moth Damage to Pine Terminal, Ron Billings, Texas Forest Service.

Insecticide spraying for tip moth control has not been a general practice, except in Christmas tree plantations, seed orchards, forest nurseries, and research and progeny tests. However, tip moth control is increasingly becoming a component of intensive short-rotation pine plantation management. Effective control requires that insecticide applications coincide with egg hatch and larvae emergence. Trapping or monitoring moth emergence to predict egg hatch and larval development is the critical element in a tip moth control programme.

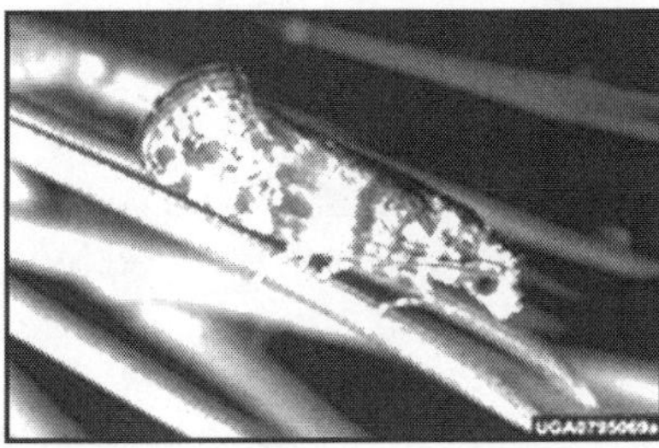

Fig. Nantucket Pine Tip Moth on Loblolly Pine Needle, James A. Richmond, USDA Forest Service.

WEEVILS

The snout beetles (Coleoptera: Curculionidae) are a diverse and abundant group of insects. Many weevils are important destructive pests of agricultural, horticultural and forest crops. Weevil larvae are creamy-white and legless, with brown head capsules. Adults are hard-bodied, cylindrical beetles with a pronounced "snout" that contains hardened, chewing mouth parts at the end.

The two most common species of boring pine weevils in the South are the white pine weevil, *Pissodes strobi* (Peck), and the deodar weevil *Pissodes nemorensis* Germar. Weevils in the genus *Curculio* attack the seeds of nut-bearing trees, most notably acorns, but several species are important pests on hickory, chestnut, and pecan. During certain years, the nut crop can be almost completely destroyed. In addition to the species listed, there are many other species of weevils that are important in forest environments.

WHITE PINE WEEVIL

The white pine weevil is the most serious pest of eastern white pines in the South. This weevil can also feed on and reproduce on a variety of spruce and pine species. Adult weevils are 1/6 - ¼ inch long, brownish and marked with irregular gray-white patches. Adults overwinter in litter under the trees. The adults emerge from hibernation in the spring and begin feeding on cambial tissues of the main stems of the host plant, usually within 1 inch of the terminal bud. The female lays eggs in feeding pits on the terminals. After hatching, the larvae tunnel downward in the cambium. This feeding girdles and kills the leader. Mature larvae pupate in chambers formed in the wood. Adults emerge throughout the summer. Only one generation per year has been reported.

Fig. White pine weevil adult, E. Bradford Walker, Vermont Dept. of Forests, Parks, and Recreation.

The first sign of white pine weevil attack is pitch flow from feeding punctures on the terminal shoots. Later the new growth appears stunted, and finally, the needles wilt and the terminal dies. Trees up to 3-4 feet tall may be killed. Dead terminals on larger trees are replaced by one or more branches of the topmost living whorl, resulting in crooked or forked stems.

Management practices for the white pine weevil include:

- Mixed planting of white pines with hardwoods or planting white pines under hardwood cover;
- Planting only on soils where the hardpan is three or more feet below the surface;
- Selecting and pruning the least injured pines in preparation for later harvest; and
- Removal of less desirable pines from damaged stands.

Drenching susceptible trees with pesticide sprays as adult weevils emerge from hibernation can provide some protection. Insecticide sprays are used to protect white pines grown for Christmas trees from damage.

Fig. White Pine Weevil Damage to Terminal, E. Bradford Walker, Vermont Dept. of Forests, Parks, and Recreation.

DEODAR WEEVIL

The deodar weevil attacks most species of pines and many introduced cedars (Figures). These weevils cause damage by feeding on young shoots in the crowns of sapling and pole-sized trees. Both adults and larvae kill terminals and cause branch-end flagging on pole-sized and small sawlog size trees. Adults are active and lay eggs all winter. The adult weevil is about ¼ inch long, grayish-brown to dark brown with whitish spots on the wing covers. Adults are attracted to weakened, stressed or dying trees. They often breed in logging slash and trees killed by bark beetles. Adults chew holes and feed on the inner bark and wood of twigs and leading terminals. After chewing through the bark, females deposit eggs in the inner bark of host trees. Following egg hatch, the larvae feed beneath the bark much like white pine weevils, girdling and often killing the stem. Evidence of their presence is indicated by swelling of the bark over feeding areas. Pupation occurs in chip cocoons in the sapwood beneath the bark. Adults apparently become inactive (aestivate) during the summer months but appear again in the fall to feed on twigs and leading shoots. May is the month of greatest adult emergence.

Fig. Deodar Weevil Pupa in Chip Cocoon, Gerald J. Lenhard, LSU.

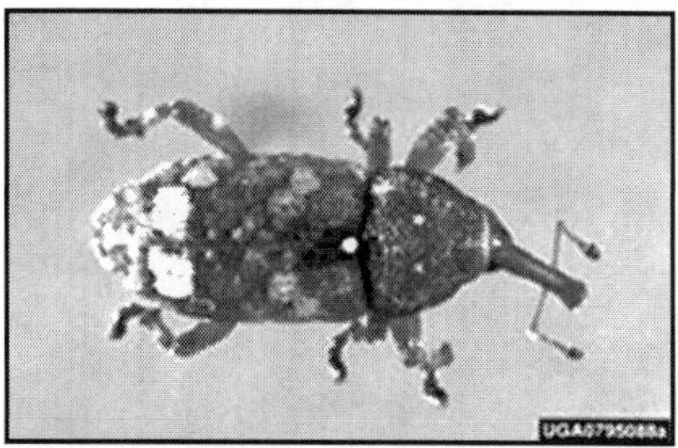

Fig. Deodar Weevil Adult, Gerald J. Lenhard, LSU.

CHEWING INSECTS

This grouping, based upon USDA Forest Service Pest Trend Impact Plot System, feeds on stems and shoots but does not include the leaf-eating insects, or borers. The more important members of this group in the South are the reproduction weevils which will be covered here.

REPRODUCTION WEEVILS

The pales weevil, Hylobius pales (Herbst), and the pitch-eating weevil, Pachylobius picivorus (Germar), are very destructive pests of young pines

(Figures). They feed and develop on all species of pines within their range. Adult pales weevils are ¼ - 1/3 inch long with patches of yellow hairs appearing as bars across the wing covers. Pitch-eating weevils are slightly larger (1/3 - ½ inch) with yellowish spots on the wing covers. They spend the winter as adults in the soil. Overwintering adults emerge during the spring and feed on the bark of saplings and at the bases of seedlings. Most damage occurs in the spring and fall. These weevils feed at night and hide in the soil around the base of seedlings during the day. After feeding, females lay eggs on roots of recently cut, damaged or killed pines. Larvae burrow and feed on root tissue and later pupate in chip cocoons under the bark.

Fig. Pales Weevil Adult, Wayne N. Dixon, Florida Dept. of Agriculture and Consumer Services.

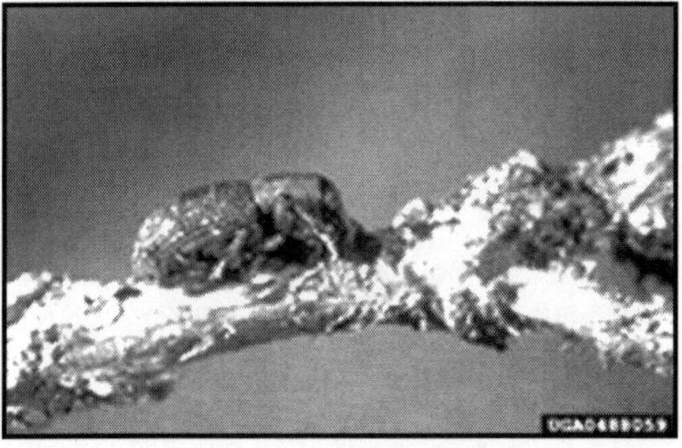

Fig. Pitch-eating Weevil Adult and Feeding Damage, Robert L. Anderson, USDA Forest Service.

Damage or death of pine seedlings often occurs when adults of these weevils eat patches of bark from the stems. When feeding areas overlap, the seedling is girdled. Christmas tree plantations are sometimes seriously damaged by these weevils. The most practical and economical method of controlling damage is to delay planting in areas harvested after June for at least one year. If planting cannot be delayed, chemically control reproduction weevils by root dipping seedlings in an insecticide and kaolin clay mixture, top dipping of seedlings, or over-the-top spraying of seedlings prior to lifting in the nursery or after transplanting.

Fig. Pitch-eating Weevil Adult Feeding, Wayne N. Dixon, Florida Dept. of Agriculture and Consumer Services.

DEFOLIATING INSECTS

Included in this group are insects that eat leaves and needles. There is a diverse and broad array of insects that includes the many caterpillars, sawflies, leafcutting wasps, bees and ants, beetles and walkingsticks. Trees attacked by defoliators can be recognized by missing foliage and uneaten leaf parts such as veins and petioles.

Additionally, many members of this group feed within a leaf, mining between the upper and lower epidermis. Defoliation reduces photosynthesis, interferes with transpiration and translocation within the tree. Light defoliation normally has little affect on the tree, but moderate-to-heavy or repeated defoliation can reduce tree vigour. The impact that defoliation has on a tree depends upon the time of the year, the tree species, tree health, and whether defoliation occurs more than one time. It is important that the manager properly identifies the organism(s) involved and clearly understands the dynamics of both the forest stand and the insect involved before any management scenario is developed and implemented.

GYPSY MOTH

When it is present, the European gypsy moth, *Lymantria dispar* (Linneaus), is one of the most destructive hardwood forest pests. A native of Europe, the gypsy moth was accidentally introduced into the U.S. in New England in the late 1800s and has gradually spread into at least 17 eastern states, as well as into other parts of the U.S. and Canada. The generally infested area now includes all of the northeastern states and portions of West Virginia, Virginia, Michigan and Ohio.

Between 1982 and 1996 in the U.S., gypsy moth defoliation ranged from less than 1 to more than 8 million acres per year. Female European gypsy moths are incapable of flight but are prolific egg layers. Females lay eggs in masses in trees and on lawnmowers, outdoor furniture, mobile homes, recreational vehicles, firewood, building materials, doghouses, and other items left outdoors. New gypsy moth infestations occur through inadvertent transport of egg masses and pupae. Local infestations can spread as small larvae move from one site to another on air currents for distances of a few feet

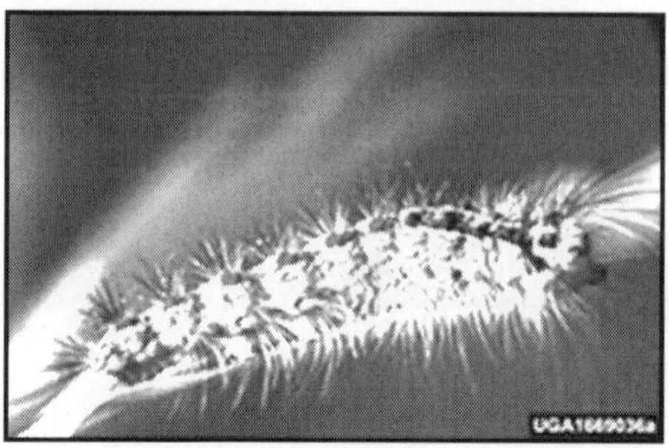

Fig. Gypsy Moth Larva, James A. Copony, Virginia Dept. of Forestry to Several miles.

Fig. Gypsy Moth Females and Egg Masses, John H. Ghent, USDA Forest Service.

Fig. Gypsy Moth Defoliation, Mark Robinson, USDA Forest Service.

The female gypsy moth is heavy bodied, almost white with a wingspan of about 2 inches. The male is dark brown, with blackish bands across the forewings, and has a wingspread of about 1 ½ inches. Full-grown larvae are from 1 ½ to 2 ½ inches long. Older larvae have yellow markings on the head, a brownish-gray body with tufts of hair on each segment, and a double row of five pairs of blue spots followed by a double row of six pairs of red spots on the back. Moths are harmless, but the caterpillars from which they develop are voracious leaf feeders of forest, shade, ornamental and fruit trees and shrubs. Large numbers of caterpillars can completely defoliate an area. A single defoliation can kill some softwoods, but it usually takes two or more defoliations to kill hardwoods. Large infestations contain millions of caterpillars and can degrade aesthetic and recreational values of forests, parks and wooded homesites. The number of trees killed as a direct result of gypsy moth defoliation is relatively small, but many trees are weakened and become susceptible to secondary attack by other insects or plant diseases.

A U.S. Department of Agriculture Federal Domestic Quarantine (7CFR 301.45 gypsy moth) regulates transport of firewood, lumber, and many other outdoor items from infested areas to non-infested areas to prevent or reduce the likelihood of gypsy moth transport.

A nationwide, cooperative state-federal monitoring programme based on the use of large numbers of pheromone traps continues to monitor for accidental introductions of gypsy moths. The pheromone traps are effective in capturing the highly mobile adult males and are a good tool to monitor low-level gypsy moth populations. Control of newly developed "spots" detected by this monitoring programme in southern states in recent years has been effective. Without this regulatory action, the gypsy moth would undoubtably infest an area much larger than the current areas.

Despite the quarantine efforts, isolated infestations of gypsy moth have occurred in the southeast in North Carolina, Tennessee, Arkansas, and Georgia. As these isolated infestations were found, comprehensive eradication projects, as mandated by federal law, have been undertaken. During the past 80 years, many people have tried to control gypsy moth populations by introducing parasites into infested areas with limited success.

Ongoing research on the use of viral and fungal diseases show promise for controlling the gypsy moth. In particular, a fungus introduced on several occasions over the last 80 or so years, appears to have increased in virulence or otherwise become widespread and extremely effective on reducing gypsy moth populations in recent years. This fungus, Entomophaga maimaga, has been reported to have significantly reduced gypsy moth populations in much of New England and Pennsylvania. We can only hope that the impact of this fungus continues to increase.

Fig. Spraying Bt. by Helicopter for Gypsy Moth, G. Keith Douce, UGA.

Aerial spray programmes of an approved insecticide are still an important component of gypsy moth control programmes in infested areas. Large acreages of forests and urban ornamental trees in infested areas are treated aerially with chemical and biological insecticides each year to reduce potential damage by this pest.

SAWFLIES

Several species of sawflies (Hymenoptera: various families) can be serious defoliators of conifers in both forest and plantation stands. Sawfly adults are small broad-waisted wasps. Larvae resemble caterpillars but are usually without hairs and have five or more pairs of fleshy prolegs under their abdomen (caterpillars normally have four or fewer pairs). Larvae of the more commonly found sawflies vary from 2/3 to 1 ¼ inches long, are usually greenish to dusky gray, and have conspicuous stripes or spots. Outbreaks occur periodically, sometimes over large areas, and can result in loss of tree growth and sometimes tree mortality.

REDHEADED PINE SAWFLY

The redheaded pine sawfly, Neo-diprion lecontei (Fitch), is one of the more commonly found and most destructive sawflies in the Southeast. Red headed pine sawfly larvae are usually found on trees from 1-15 feet tall, where they feed gregariously on old and new needles and on tender shoots of these young trees.

Full-grown, redheaded pine sawfly larvae are about ¾ to 1 ¼ inches in length; have a reddish head capsule and a yellowish-white body marked with six rows of black spots. Many times a naturally occurring virus causes collapse of an infestation. Occasionally, controls are warranted during heavy population peaks. Depending upon the size of the infestation, these treatments may be applied aerially or by ground equipment.

Fig. Redheaded Pine Sawfly Larvae, Gerald J. Lenhard, LSU.

Fig. Redheaded Pine Sawfly Adult Female Oviposition, James McGraw, NC State University.

PINE WEBWORM

The pine webworm, Tetralopha robustella Zeller, may become a problem in pine plantations (Figure). The adult moth has about a ¾-inch wingspan. The basal part of the forewing is purple-black, the central part grayish, and the outer part blackish. Full-grown larvae are yellowish brown, with two dark brown longitudinal stripes on each side and are about ¾ inch long. Pine webworms overwinter as pupae in the soil. Adults emerge in late spring to early summer and deposit eggs on the needles. Young larvae mine needles, while older larvae live in silken tubes that extend through webs of globular masses of brown, coarse frass. These webbing masses enclose the needles upon which the larvae feed. At first, the webbing masses may be only one or two inches long. The webbing mass may contain several larvae and increases in size as the larvae mature. Seedlings up to two feet tall can be completely defoliated. Infestations on larger trees can cause partial defoliation resulting in loss of growth and poor tree appearance.

Fig. Pine Webworm Damage, Robert L. Anderson, USDA Forest Service.

Usually no controls are necessary unless extremely heavy populations are encountered or individual specimen trees are involved. Individual infestations can be destroyed by hand. If controls are required, the larvae are easily controlled by labelled insecticides, but good spray coverage and pressure are needed to penetrate the webs.

FALL WEBWORM

The fall webworm, Hyphantria cunea (Drury), can have two or more generations per year. Webworms enclose leaves and small branches in their light gray, silken webs. Fall webworm is known to feed on more that 100 species of forest and shade trees. In the eastern U.S., pecan, walnut, American elm, hickory, fruit trees, and some maples are preferred hosts. The moth is white with dark wing spots and has a wingspan of between 1.4-1.7 inches. Though the webs are unsightly, damage to most trees is considered to be insignificant and is usually of only minor economic importance as a forest pest. However, in areas where heavy defoliation occurs, including in pecan production areas, control measures may be needed.

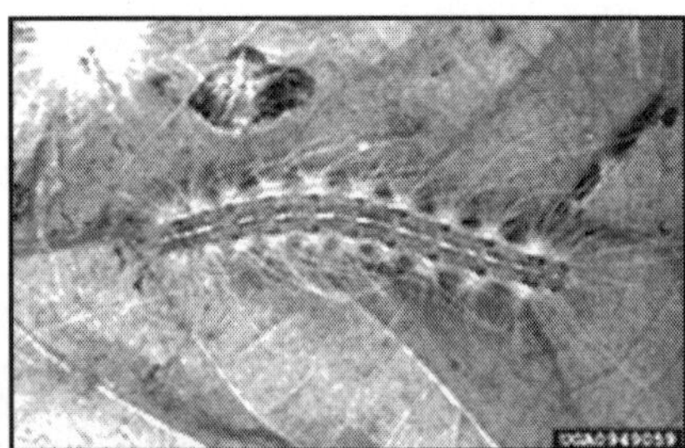

Fig. Fall webworm Larva, James B. Hanson, USDA Forest Service.

Fig. Fall Webworm Webbing, Ron Billings, Texas Forest Service.

OAKWORMS

The three common species of oakworm found in the South are the orangestriped, Anisota senatoria (J.E. Smith); the pinkstriped, A. virginiensis (Drury); and the spiny, A. stigma (Fabricius). Oakworms occur throughout the eastern U.S., are voracious feeders, and when abundant quickly strip trees of their foliage. However, since defoliation usually occurs late in the summer or into the fall, their economic impact is relatively minor. Orangestriped oakworm larvae are black with eight narrow yellow stripes; pinkstriped oakworms are greenish-brown with four pink stripes; and the spiny oakworm is tawny with pinkish short spines. Larvae have a distinctive pair of long, curved "horns" on the dorsum behind the head.

EASTERN TENT CATERPILLAR

The eastern tent caterpillar, Malascoma americanum (F.), is primarily an aesthetic problem (Figures). Species of the genus Prunus are preferred hosts, with black cherry being the preferred non-cultivated host. Full-grown larvae are between 2 to 2 ½ inches in length, have black heads and long light-brown body hairs.

The back has a light stripe bordered on each side with yellowish-brown and black wavy lines. The sides are marked with blue and black spots. Eastern tent caterpillar overwinter as eggs in shiny, dark brown masses around small limbs on host trees. Eggs hatch in early spring, and the larvae begin to construct a tent and enlarge the structure as they feed and grow. Chemical controls are usually not justified. Defoliated trees normally refoliate and suffer only minor growth loss.

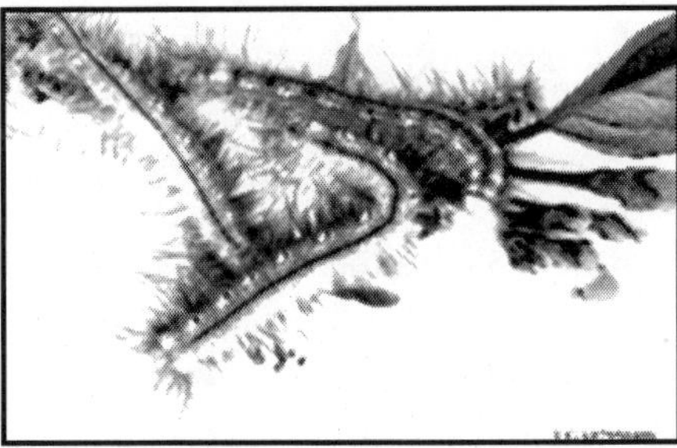

Fig. Eastern Tent Caterpillar Larvae, Gerald J. Lenhard, LSU.

Fig. Eastern Tent Caterpillar Tent in Tree Branches, Robert L. Anderson, USDA Forest Service.

There are numerous other defoliating insects that will not be covered here.

SUCKING INSECTS

This group of insects has piercing-sucking mouthparts which they use to pierce plant tissues and suck sap from the plant. Insects in this group that attack trees are in the Orders Homoptera and Heteroptera. In addition to these insects, many species of mites in the Class Arachnida: Order Acari also feed on plants. Only a few species of sucking insects kill forest trees directly.

Among the observable symptoms of feeding by sucking insects are:

- Discoloration of needles or leaves,
- Curled foliage,
- Honeydew and sooty mold on leaves, stems, twigs and other materials,
- Fine silk webbing on the needles and leaves,
- Premature leaf drop,
- Branch mortality,
- Oviposition scars made by cicadas and treehoppers, and
- Galls.

In addition to their direct feeding damage, some sucking insects are vectors of plant diseases. Many of these insects are individually quite small and are frequently transported on nursery stock. As before, we will only discuss a few of the major forest pest species that make up this complex group.

SCALE INSECTS

A number of species of these small sucking insects are important in forest environments (Figure). Adult females lack wings, may not have legs, and are saclike with no definite body segmentation. Adult males are more insect in appearance, usually with one pair of wings and with a definite head, thorax and abdomen. Most scale insects produce a waxy substance that covers the body either as a shield-like structure or as a coating on the body surface. Natural dispersion is usually by windborne, first instar “crawlers” that are equipped with legs and can be quite mobile. In most cases, instars other than crawlers are generally sessile (do not move). The small size and cryptic appearance of many scale insects has greatly helped disperse many scales inadvertently as contaminants on plants transported during commerce. Scale insects damage plants by inserting their sucking mouthparts in plant tissue and ingesting large amounts of plant sap. Scale insects also excrete large amounts of honeydew which serves as a substrate for the growth of sooty mold. Plant deformation and toxin injury are produced by some species of scale insects. Natural enemies are frequently important in regulating scale insect populations. Scale insects occasionally become a problem in pine seed orchards where other pesticide applications have eliminated or reduced natural predator and parasite populations. Important groups of scale insects in forestry include mealy bugs, soft scales, armoured scales and Kermes scales.

Fig. Pine Needle Scale Infestation on Foliage, James B. Hanson, USDA Forest Service.

The scale life stage most susceptible to chemical control is the first instar crawler stage. During this part of their life cycle, there is little or no waxy covering on the body. Attempts to control scale insects during other life stages are greatly hampered by the waxy covering on the insect's body and are often not very successful. To achieve control, monitor crawler emergence and time control efforts accordingly.

APHIDS

Aphids are common pests on trees throughout the South. All southern pines are subject to aphid attack. They are soft-bodied, usually wingless insects less than 1/8 inch long. They may be pink, brown, black, whitish or greenish. The rate of development and reproduction of aphids is very rapid, producing many generations each year.

Aphids suck plant juices from the tender, succulent parts of plants. Heavy feeding causes stunting of terminal buds, and needles become distorted or stunted. Often the first sign of attack is the presence of many aphids on the branches. They excrete sweet, sticky honeydew which may attract many ants. A fungus, sooty mold frequently develops on the honeydew and the branch or entire tree may appear black. Automobiles parked under heavily infested trees will frequently be covered with this sticky honeydew.

Chemical control of aphids is generally not economically justified except in cases of very high value or aesthetically important cases, such as Christmas tree plantations or nursery stock. Aphids can be controlled by an application of labelled insecticides when necessary.

BALSAM WOOLLY ADELGID

The balsam woolly adelgid was accidentally introduced into the U.S. from Europe around 1900. It has become a serious pest of natural Fraser fir stands in the southern Appalachians and thus causes considerable damage to the Fraser fir Christmas tree industry. The impact of this adelgid has been severe. Complete stand mortality, severe timber losses and reduced tree growth have been observed. This insect has killed millions of board feet of true fir timber in North America. Adults are blackish purple, roughly spherical in shape, and about

1/32 inch in length. The insect produces a covering of white wax threads on the surface of the tree's bole, limbs and buds. In the South, there are two to three generations of the adelgid per year. Orange-coloured eggs are produced and remain under the adult's body until hatching. The newly hatched "crawler" is the only stage of the adelgid that is mobile. When the crawler begins feeding, it transforms into a first instar nymph and becomes stationary. During the feeding process, the host tree is stimulated to produce abnormal wood that reduces the trees ability to translocate food and water. A heavily infested tree may die within 2 to 7 years. Chemical controls can be quite effective but are extremely costly and are usually limited to high value trees.

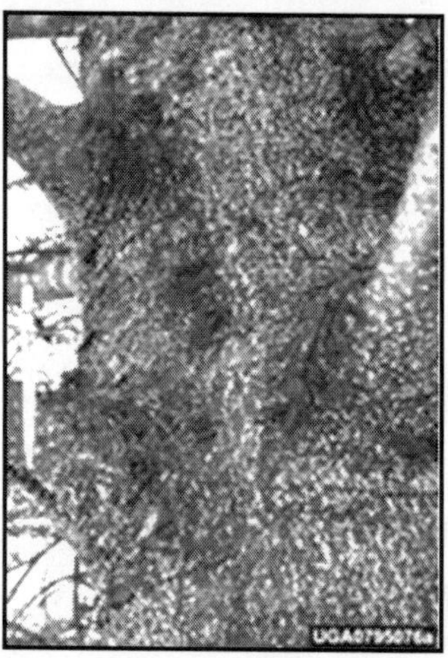

Fig. Balsam Woolly Adelgid Infestation, USDA Forest Service Archives.

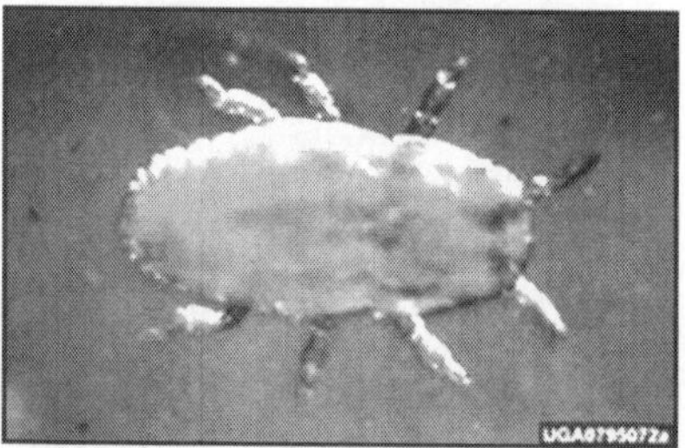

Fig. Balsam Woolly Adelgid Infestation, USDA Forest Service Archives.

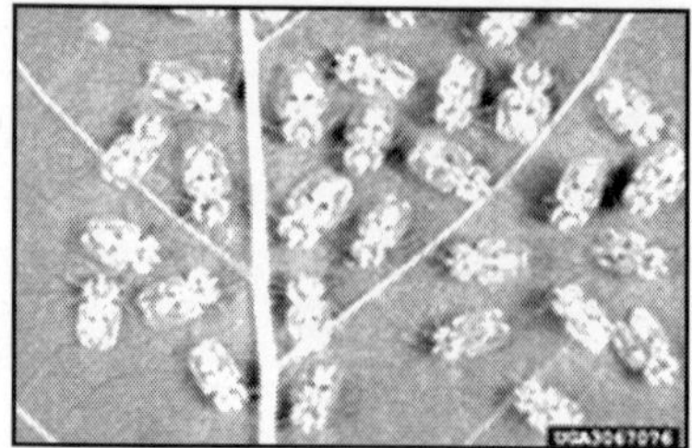

Fig. Sycamore Lace Bug Adults.

LACE BUGS

Lace bugs, Corythucha spp., feed on the leaves of many tree species. Both the adults and nymphs feed on leaves, often resulting in chlorotic flecks or tiny chlorotic spots on the upper leaf surface. In addition to the presence of numerous nymphs and adults, the underside of leaves upon which lace bugs are feeding usually has numerous cast nymphal skins and numerous small black "frass spots" and black fungus. Heavily infested trees may be partially or fully defoliated, especially in dry weather. There may be several generations per year, and all life stages reside on the leaves of the host tree. Both the nymphs and the adults feed by inserting their mouthparts into the leaf tissue and sucking plant juices. Nymphs are dark-coloured and covered with spines. Adults have broad, transparent, lacelike wing-covers. The adults are flattened and are about ¼ inch in length. Natural enemies are usually effective in controlling populations. Chemical controls are usually only used on high value shade and ornamental trees.

SPIDER MITES

Spider mites are found throughout the South. A number of species are important pests of ornamentals and shade trees, as well as many other plants. Spider mites are less than ½ inches in length, and, depending upon the species, vary in colour from yellowish, greenish, orangish, and reddish to red. Major symptoms of spider mite damage are silk webbing, cast skins, active mites, and discolored yellowish foliage. Spider mites spin very fine silk webbing as they move about. There are several generations of mites per year. During high infestations, infested foliage may be discolored, disfigured, or killed. Insects in Seed Orchards and Forest Nurseries

Fig. Spider mite damage, Robert L. Anderson, USDA Forest Service.

High value, intensively managed sites, such as seed orchards and forest nurseries, require aggressive forest insect control programmes. A number of insects that are normally not considered economic forest pests can be quite damaging in seed orchards and forest nurseries. Pheromone traps are often used to monitor insect populations in these sites. Some of the major seed and cone insect pests are the southern pine coneworms, pine seedbugs, various sawflies and thrips.

CONEWORMS

Several species of coneworms (Dioryctria spp.) are highly injurious to seeds and cones of conifers. These insects infest all commercially significant pines as well as spruce, fir, hemlock, and cypress. The southern pine coneworm (D. amatella) infests cones, male flowers, shoots, and fusiform rust cankers on a variety of southern pines.

Adults have a wingspan of about 1 and 1/8 inches. The forewing is dark brown with contrasting white patches in zigzag lines running across the wings. Mature larvae are brownish to purplish above, pale whitish to greenish below and about one inch long. This species is frequently reported to cause heavy cone losses to southern pines.

Fig. Southern Pine Coneworm larva, adult, and damage, R. Scott Cameron, International Paper.

Fig. Webbing Coneworm Adult, Larry R. Barber, USDA Forest Service.

The Heteroptera, the true bugs, contains two families that are significant pests of a number of conifer species. The coreid bugs (Coreidae) and the stink bugs (Pentatomidae) feed on the ovules and seeds of pines and conifers.

LEAFFOOTED BUGS

Leaffooted bugs in the family Coreidae genus Leptoglossus are important pests of loblolly and shortleaf pines. Both the nymphs and adults are reddish-brown to gray and have long legs with a laterally expanded "leaflike" tibia on the hind leg. The adults are 2/3 - ¾ inches in length and have distinctive whitish marks across the wings.

There are several generations produced each year. Nymphs and adults have piercing-sucking mouthparts that they insert into the conelets or cones to penetrate and feed upon the developing ovules and seeds. Attacked cones show no external damage symptoms, but damage to seed can be severe.

Fig. Leaffooted Pine Seed Bug Adult, R. Scott Cameron, International Paper.

STINK BUGS

Stink bugs in the family Pentatomidae emit a disagreeable odour when they are disturbed. The shieldbacked pine seed bug, Tetyra bipunctate (Herrich-Schaffer), is an important pest in southern pine seed orchards. The adults and nymphs are oval and have a humpbacked appearance. The adults are about 2/3 inch in length and are gray-brown to reddish-brown in colour. There is only one generation per year. Nymphs and adults both have piercing-sucking mouthparts which they insert into cones to penetrate the seeds. Most of the damage occurs in late summer and fall, which results in poor seed viability and low yields of sound seeds.

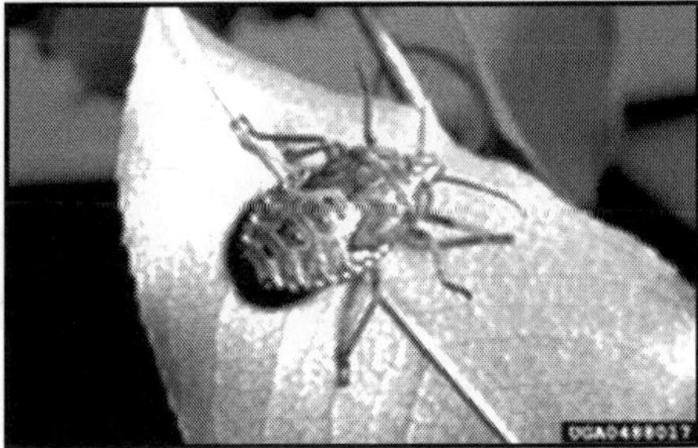

Fig. Stink bug nymph, John H. Ghent, USDA Forest Service.

A number of insecticides are specifically labelled for and are used in seed orchards and forest nurseries. Ground-based hydraulic sprayers, airblast sprayers (mistblowers), and handheld compressed-air sprayers can be used to apply pesticides. Aerial applications are made with both helicopter and fixed-wing aircraft, equipped with either conventional or by ultra-low volume equipment.

GALLMAKERS

When feeding on plant tissues, many insects and mites inject or secrete a substance into the plant that causes the plant to grow abnormal "galls" (Figure). Galls may be found on leaves, buds, stems, or roots. Plant galls are caused by a number of different animal and disease organisms, but the majority are caused by insects and mites. The greatest majority of galls are produced by cynipid wasps (Family Cynipidae), gall midges (Family Ceccidomyiidae), and eriophyid mites (Class Arachnida: Order Acari: Family Eriophyidae). However, other wasps, mites, flies, beetles, homopterans, and lepidopterans also produce galls. Each species of insect or mite produces a characteristic gall on a certain part of a specific plant. The host involved, the location, and the shape of the gall produced are extremely useful in identification of the causing species, since the actual organism itself is small or may have already vacated the gall.

Fig. Typical oak Gall, Robert L. Anderson, USDA Forest Service.

Insect and mite galls are not considered economically important in forest stands. Gall makers are considered important pests on certain ornamental trees and shrubs and on some Christmas trees. Usually controls are not needed for gall makers but may be desired in some situations such as Christmas tree plantations and ornamental plantings.

DISEASES

ROOT/BUTT DISEASES

Pathogens that attack root systems may affect small feeder roots as with Littleleaf disease, or a pathogen like Annosum root disease can attack larger roots causing decay extending into the butt of the tree where lateral roots attach to the trunk. In general, root diseases are more prevalent on sites which have been altered by erosion, compaction, imperfect drainage, or other disturbances.

Along with site factors, environmental stress, host characteristics, and interactions with other microorganisms lead to root disease complexes which may become prevalent in plantations.

LITTLELEAF DISEASE

Littleleaf is the most serious disease of shortleaf pine in the South. However, in localized areas, loblolly pine can also be severely affected. Other pine species are much less susceptible to littleleaf disease. As the name implies, littleleaf disease results in shortened, stunted yellow needles. Early symptoms are difficult to distinguish from nutrient and water deficiencies. As the disease progresses, the foliage is thinned; tufts of needles only remain on branch terminals; and twigs and branches die throughout the crown. Normal needle length of 3-5 inches is reduced to about 2 inch. Although cones may be abundantly produced, they will be very small and contain only a few viable seeds. Littleleaf disease is caused by a combination of poor soil-water drainage and subsequent attack of the feeder roots of pine by the fungus Phytophthora cinnamomi. The P. cinnamomi fungus is found throughout the growing range of shortleaf pines. The fungus requires free moisture to survive and reproduce. As new root-tips and very young feeder-roots are attacked and killed by the fungus, nitrogen uptake is reduced, resulting in slowed growth and yellowing of the foliage. Relationships between poor soil drainage and root infection confine the disease to heavier, fine-textured soils.

Fig. Littleleaf Disease on Pine, USDA Forest Service Archives.

Fig. Littleleaf Disease Typical Soil Profile of Susceptible Site, USDA Forest Service Archives.

Stands are seldom affected prior to 20 years of age. The manifestation of the disease is most intense in stands more than 40 years old. Infected trees may survive for up to six years after first visible symptoms appear, although a few may die the first year. Infected trees rarely recover from the effects of the disease unless they can be fertilized, which is generally only practical in an urban situation.

Do not plant shortleaf pines on soils with poor internal drainage. The lack of oxygen in these soils prohibits regeneration of roots and restricts root development into the soil. When the available nitrogen is depleted in P. cinnamomi infested soil, littleleaf disease develops. Favour more resistant loblolly, slash and longleaf pine over shortleaf pine in areas susceptible to littleleaf disease.

ANNOSUM ROOT DISEASE

Caused by the fungus Heterobasidion annosum, annosum root disease can infect all pine species in the South. Fungus fruiting structures (conks) may appear on the bark surface at the root collar area of the tree or stump. Spores liberated from these conks can cause local and long-distance spread of the fungus. Spores, which are air-blown to the surface of freshly cut stumps, germinate rapidly and infect the stump. Mycelium of the fungus grows into the stump root system and is transmitted through root contacts or root grafts to roots of healthy trees. Spores of the fungus have the ability to infect directly and are more prominent on stump roots than on roots of healthy trees. Roots damaged and exposed when plowing fire breaks or during road construction are susceptible to becoming infected as well.

Suspect annosum root disease if tree mortality begins the second or third year following thinning and continues for several years. Infected trees show a general lack of vigour, shortened needles and internodes, chlorosis and heavy cone production. However, infected trees frequently do not show the symptoms and may fall over before any injury is noticed. Decline can be rapid or may take several years.

Fig. Annosum Root Disease Pitch-soaked Wood and Sand, Robert L. Anderson, USDA Forest Service.

Fig. Annosum Root Disease; Windthrown Diseased Trees, Robert L. Anderson, USDA Forest Service.

Examine lateral roots for pitch-soaking and white, stringy decay of terminal considered a high-hazard site for annosum. Obtain soil survey maps to study soil for characteristics of texture and drainage. Interpret these results as indications of a low- or high-hazard site. When soil maps are not available, roadside cuts offer clues about the soil profile. When 50 per cent or more of the land area is determined to be high-hazard, the entire stand should be managed as a high-hazard site.

In high-hazard areas, thin the stand only in the summer as high summer temperatures limit spread and viability of the spores. Stumps may also be treated with borax immediately following harvest. If mortality continues for five years after the first thinning, clearcut the stand and regenerate.

STEM DECAYS/CANKERS

Gall and canker forming pathogens pose serious problems in forest management through tree loss and stem quality degrade. The most serious problems occur in southern pines, although canker diseases of hardwoods cause significant losses as well. Chestnut blight, caused by Cryphonectria (Endothia) parasitica, eliminated American chestnut in North America in less than 50 years after it was introduced from the Orient in 1900. Other frequently observed hardwood canker diseases include, most hypoxylon dieback (H. atropuntatum), nectria canker (N. galligena), strumella canker (Urnula craterium), and fusarium canker (Fusarium spp.).

These hardwood cankers seldom result in direct mortality but cause substantial stem quality degrade and increase stem breakage. Most hardwood canker diseases occur when individual tree vigour declines following environmental stress and/or mechanical damage, such as fire, wind damage, or logging damage.

FUSIFORM RUST

Fusiform rust of slash and loblolly pine causes extensive economic loss annually in pine stands. These losses are compounded by the impact fusiform rust has in nurseries and young plantations.

Fig. Fusiform Rust Gall, Robert L. Anderson, USDA Forest Service.

Fig. Fusiform Rust Infected Seedlings, Robert L. Anderson, USDA Forest Service.

The most easily recognized symptom is the spindle shaped canker on pine branches or main stems. In early spring these swellings appear yellow to orange as the fungus produces powdery spores. Older stem cankers may become flat or sunken as host tissue is killed. Cankers often girdle trees; wind breakage at the canker is common. Spores of the fungus Cronartium quercum f. sp. fusiforme produced on pine infect oak leaves. Brown, hairlike structures are produced on the underside of the oak leaves in late spring. Spores are produced, which in turn reinfect pine trees, completing a typical rust cycle.

Disease control in pine stands is primarily achieved by removing infected trees during thinnings and using genetically superior seedlings. On high-hazard sites, plant resistant species using locally or regionally improved selections. Plant pines within their natural range, especially slash pine. Plant only disease-free nursery stock.

On bare, high-hazard sites, use a minimal level of site preparation to give satisfactory plant survival and growth. Destroy present host oak populations by burning, herbicide treatment or girding large residual oaks. Prevent resprouting of oaks. Limit the size of planting blocks to increase variation in age classes and plant material. Planting density can be adjusted to compensate for random rust infection and mortality but only to the extent that growth and yield are not affected before scheduled thinning.

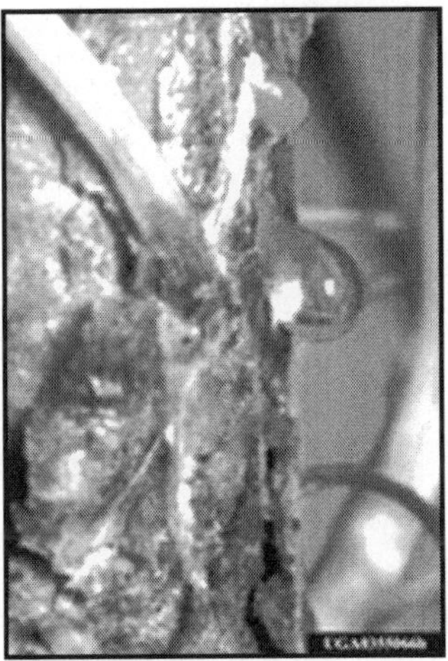

Fig. Fusiform Rust Pycnia, Robert L. Anderson, USDA Forest Service.

Delay fertilization of slash and loblolly until the 10th year in regions of moderate or high-hazard. In low-hazard, flat-wood sites, growth response of slash pine to fertilizer will offset the rust impact.

When the stand is 3-5 years old, estimate rust incidence to determine future management policy. If disease incidence is high, consider sacrificing the stand early to replant with resistant plant material. Burning will help keep intruding oak hosts under control. Potential losses in older, heavily infected stands can be reduced by presalvage harvesting of rust infected trees. Burn after harvesting.

Nurseries should be located in low-hazard areas. Host oaks and infected pines should be eradicated from the area before establishing a nursery. Apply protective fungicides according to the label. Since warm, moist conditions are required for fungus infection, irrigate only during the middle of the day during the rust season. This allows the foliage to dry, rather than remain wet overnight.

PITCH CANKER

The pitch canker fungus, Fusarium subglutinans (Wollenweb. and Reinking), causes growth loss and mortality to many pine species including Virginia, slash, loblolly, shortleaf and longleaf. Pitch canker is characterized by copious pitch flow and pitch-soaked wood. Shoot cankers result in dieback, characterized by wilting and killing of the crown. Needles on infected shoots turn yellow to reddish-brown, later turning greenish-brown to dark gray.

Fig. Pitch Canker: Stem Canker, Robert L. Anderson, USDA Forest Service.

Fig. Pitch Canker Fungus Growing from Seed, Robert L. Anderson, USDA Forest Service.

Cankers on trunks and large limbs are perennial, while cankers on shoots are usually annual. Infected pole-size trees usually have annual shoot cankers and may die from extensive infections. Less affected and younger trees may not be killed but may suffer reduced growth and loss form. Infected seedlings exhibit yellow-green to reddish-brown needles and wilting foliage. Pitch-soaked lesions occur at and just above the soil line. Infected seedlings will inoculate healthy stock when inter-mixed during handling and transit. Take care to cull out pitch canker infected stock during transplanting. Infected seedlings are usually killed by the disease. Systematic removal of infected trees reduces inoculum sources and fire hazard, as well as providing growth space for other trees.

FOLIAGE DISEASES

Maintaining optimum leaf area is critical to tree health, growth and yield. Premature shedding of infected foliage is a defence mechanism which isolates and removes the attacking pathogen. However, repeated loss of effective photosynthetic leaf area from foliage diseases reduces growth and yield, predisposing trees to damage by secondary agents such as insects and environmental stress.

HARDWOOD ANTHRACNOSE

Numerous hardwoods are susceptible to anthracnose (Apiognomonia spp.) which may only cause lesions on foliage or may invade and kill leaves, twigs, and branches. Anthracnose is common throughout the South and is most prevalent on sycamore and oak. Initial infection occurs in the spring during cool wet periods following leaf emergence, most commonly from mycelium that over-wintered in previously infected twigs on the host tree. Severe infections occur when temperatures are below 50° F for two weeks following leaf emergence. The infection enters the leaf and may grow into the leaf petiole and twig; and as twigs are girdled, dieback occurs. A canker may form which will girdle twigs in the following years. Necrotic areas develop along the veins and midrib of the leaves. Deformed blighted leaves may remain on the tree, but generally they are shed as the infection spreads over the leaf. As leaves, twigs and shoots die, complete crown defoliation can occur in the spring, but this is generally followed by a second leaf flush. Secondary infections can occur throughout the growing season when moisture is present. In intensively managed sycamore plantations, use wide spacing to increase airflow between trees to aid in drying to reduce secondary disease cycles. Fungicide sprays may be applied at bud break and during early leaf development to provide protection for trees in plantations and ornamental settings.

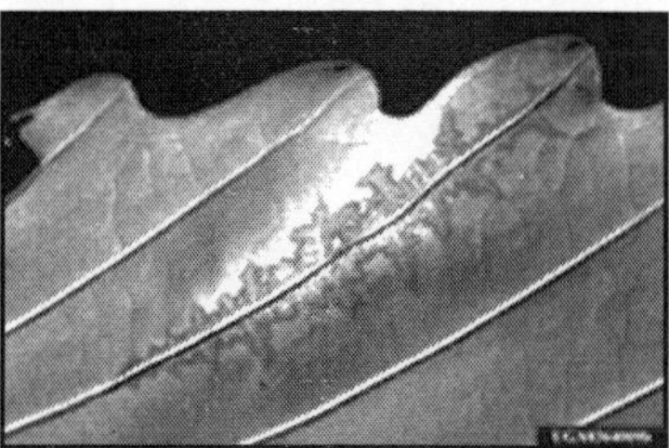

Fig. Oak Anthracnose Leaf Symptoms, Robert L. Anderson, USDA Forest Service.

Fig. Sycamore Anthracnose Leaf Blight, Robert L. Anderson, USDA Forest Service.

NEEDLECAST

Needlecast (Lophodermella spp.) is a very common disease of conifers throughout the southern United States. The disease rarely causes significant economic impact on forest trees although there is undoubtedly some reduction in growth associated with premature loss (cast) of foliage (Figure). Severe needlecast, in combination with other stresses, may contribute to vulnerability of trees to bark beetle attack.

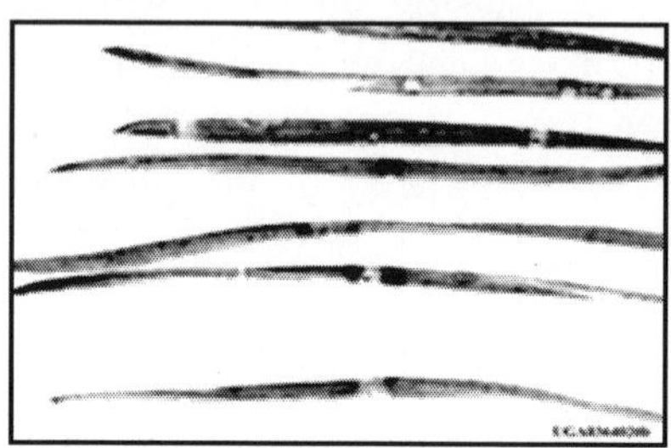

Fig. Needlecast Close-up of Fruiting Bodies, USDA Forest Service Archives.

Infected needles develop chlorotic spots beginning in winter or early spring that rapidly turn tan to reddish-brown from their tips. Characteristic black, raised football-shaped fruiting structures form on the infected needles. Thinning of the crown may result from needle drop of infected needles, leaving tufts of green uninfected needles at the branch tips. Needle cast symptoms rarely affect the entire needle. At maturity, these fruiting structures discharge spores that can infect healthy foliage.

Control is seldom feasible under forest conditions. In nurseries, shade trees and Christmas tree plantings, recommended fungicide applications may be economical.

BROWN SPOT

The most serious needle disease of longleaf pine is caused by the brown spot needle fungus, Mycosphaerella dearnessii (Scirrhia acicola). Infected needles in the early stages are irregularly yellow to brown spotted, with green tissue in between the spots. Needles are eventually killed by the girdling action of the fungus.

Fig. Brown Spot: Prescribed Burn for Control, USDA Forest Service Archives.

Longleaf pine seedlings can be seriously damaged. Severe needle blight on young seedlings can increase the length of time it takes longleaf pine to grow out of the grass stage. The disease affects both planted and natural seedlings in the field.

Brown spot can be controlled in nurseries by spraying with an approved fungicide. Prescribed burning can control the disease in longleaf pine seedling stands (Figure).

FUNGICIDES

Disease control measures and fungicide applications in most forest situations are not usually cost effective. Fungicides are recommended in nursery production to protect these high volume seedlings from infections. Seeds are generally treated with fungicides prior to planting to inhibit fusiform rust and soil-borne damping-off fungi.

VERTEBRATE PESTS

Several kinds of mammals and two kinds of birds sometimes damage living pines in the South. Their damage may vary from insignificant to serious. The mammals—deer, rabbits, squirrels, and other rodents—are the most serious pests.

Mammals prefer to feed on plant materials that have been fertilized and have a high moisture content. Periods of drought may intensify the damage of certain rodents, when they may eat bark for moisture. Some of these animals are protected as game animals, and permits are required for control harvests outside of normal hunting season and bag limits. Check state and local regulations before acting.

RABBITS

Rabbits commonly found in the southeast are cottontail rabbits, marsh rabbits and swamp rabbits. Nearly all southeastern forest habitats have at least one species of rabbit present. Rabbits prefer brushy vegetation that offers ample cover for hiding. Although rabbits are not normally destructive to well-established forest trees, they can cause considerable damage to nurseries and portions of newly planted stands by nipping off seedlings. Rabbit cuttings look different from deer browsing because the cut edges are smooth, as if done with a knife.

Deer have only upper front teeth and must pinch and pull stems, which leaves a broken end. Chemical repellents can stop rabbit damage temporarily. Thirty-inch high woven mesh fences will exclude rabbits from nurseries. Box traps and shooting (if permitted) can reduce rabbit numbers in damage areas. Removing brush piles and other cover areas may be effective in reducing high rabbit populations.

DEER

The most serious deer damage occurs from browsing on seedlings in nurseries and in young plantations. Deer frequently damage saplings by rubbing them with their antlers (Figure). This rubbing behaviour, which may remove the bark, is usually associated with the breeding season in the fall and early winter. Damage may be reduced with chemical repellents or eliminated by excluding deer with suitable fences. Shooting can reduce deer numbers in damage areas. However, deer are protected game animals, and a permit is required to shoot depredating deer when the hunting season is not in effect.

TREE SQUIRRELS

Fig. Slash Pine Rubbed by Deer, Gerard D. Hertel, USDA Forest Service.

Tree squirrels, including the gray squirrel and the fox squirrel, are known to cause damage to trees by chewing bark from trunks and branches. Fox squirrels are particularly likely to damage pines. This damage occurs sporadically and is associated with high populations of these animals. Squirrels have two breeding seasons per year.

Their populations may have periodic highs and lows not associated with losses due to hunting. Squirrels may also cause damage by feeding on pine cones in seed orchards. Intensive hunting can reduce squirrel damage in some cases where permitted.

BEAVER

Beavers are probably the most serious animal pest of timber in the Southeast. Beavers construct dams which flood forest land. They also girdle stems and fell trees (Figure). Persistent removal of beavers with appropriate traps, combined with destruction of dams, can effectively reduce beaver damage. (Before undertaking such control tactics, obtain appropriate permits.) Although beaver populations increase slowly due to their low reproductive rate (two young per adult female per year), check for beaver damage periodically and trap if necessary.

Fig. Beaver-felled Tree, James Solomon, USDA Forest Service.

COTTON RATS

Cotton rats have medium brown, grizzled fur and are about 8-10 inches long, including the tail. They are known to chew the bark from young pines up to a height of about 10 inches.

This damage is sporadic and occasionally serious in pine plantations under four years old. Since cotton rats prefer dense cover, keeping the area around the young trees clean through herbaceous weed control will help to reduce cotton rat problems.

PINE MICE

Pine mice are small, short-tailed brown mice about four inches long. Although they may occur throughout the Southeast, they are rare in many areas. However, local populations may explode and cause serious damage, especially to small trees.

They chew the bark from roots below ground and stems of saplings up to a height of about four inches.

POCKET GOPHERS

Pocket gophers have a stocky body about 7-8 inches long, a large head, and an almost naked tail. The forefeet have long, heavy claws for digging. The burrows are often marked by sand mounds at the surface.

Pocket gophers avoid heavy clay soils and wet areas. They can harm orchards by their damage to large roots. Pocket gophers are not usually a pest in forest conditions.

Fig. Yellow-bellied Sapsucker Adult and Damage, James Solomon,

VEGETATION CONTROL

Weeds are unwanted vegetation that interferes with land management objectives. They are obstacles to regeneration and optimum crop growth and development. Weeds compete with crops for moisture, nutrients and light. They can be classed as weed trees, brush, vines, and herbaceous weeds.

WEED TREES

Weed trees are undesirable hardwoods and conifers. They include deformed and defective or undersized individuals of both commercial and non-commercial species. Large weed or "wolf" trees can occupy significant growing space within a stand. Weed trees reduce the economic value of otherwise healthy, desirable trees. They affect both small and commercial size trees within a stand.

BRUSH

Brush includes shrubs, small trees and woody perennials. These prevent light from reaching tree seedlings and deprive even taller commercial species of water and nutrients. It interferes with natural regeneration or planting and can create a habitat for rabbits and rodents that may damage newly planted stands.

Over time, a build-up of brush in the understory can pose a fire hazard to the tree stand.

VINES

Vines include greenbriar, Japanese honeysuckle, wild grapes, kudzu and other plants with climbing or creeping stems. All of these grow well on good forest sites. They drag down tree branches and crowns, and compete with desirable trees for light and nutrients. Vines have vigorous sprouting habits and are some of the most difficult weeds to control.

KUDZU

Kudzu is a serious weed pest in tree plantations and natural stands and is a threat to mature as well as developing stands and all regeneration. Repeated herbicide applications are essential for control. This vine spreads so rapidly it can take over the site again in 2-3 years if a single root crown is left alive (Figure). Follow-up treatments must be made for one or more years after initial treatment. Kudzu's ability to resprout following treatment varies with the stand age, root size and plant vigour. Old stands may resprout for several years.

Fig. Kudzu Vine, Kerry Britton, USDA Forest Service.

HERBACEOUS WEEDS

Herbaceous weeds retard seedling growth in new plantations and natural stands. Tree seedlings competing with herbaceous weeds may develop poorly or die, especially in time of drought. Herbaceous weeds also create favourable cover for tree damaging animals such as mice, gophers and cotton rats. They pose the potential for loss of a new plantation by wildfire. Control herbaceous weeds with herbicides labelled for this forest use. In forest nurseries, seed orchards and Christmas tree plantings, herbaceous weed control is critical. These high-value forest crops must be free of weeds to allow for proper growth and development.

HERBICIDES

Herbicides are chemicals that kill or suppress the growth of weeds. Plants are controlled by herbicides that act on the plant's physiology. Different herbicides, concentration rates, application methods and equipment enable users to control targeted weeds without undue injury to desirable plants or the environment. Herbicides are registered for the specific forest uses and application methods for which they have been tested. Uses other than those indicated on the label are unlawful and may not provide the needed control. Off-label use can cause adverse effects to non-targeted species on- and off-site by drift or movement in soil and water. Furthermore, unauthorized use may pose a hazard to human health.

MODE OF ACTION

Herbicides may be broadly classed as contact herbicides and translocated (systemic) herbicides.

CONTACT HERBICIDES

Contact herbicides kill only the plant foliage to which they are applied. These herbicides are often non-selective, affecting most plant species whether woody or herbaceous. Their use is often referred to as "chemical mowing." Because roots and even larger woody parts are not killed, resprouting may occur, and control is often short-lived. Due to poor control, the currently labelled contact herbicides are rarely used in forestry. However, treatments to kill and dry vegetation to increase fuel loading for site preparation burning can be used.

TRANSLOCATED (SYSTEMIC) HERBICIDES

Translocated (systemic) herbicides are those that must enter and move within the plant to be effective. They move to sites where they disrupt certain physiological functions. This enables them to severely stunt or kill the plant. Most herbicides used in forestry, whether applied to foliage, soil, bark or cut surface, are of this type. Some translocated herbicides work in more than one way. Some of these may also act as contact herbicides at higher concentrations because of the petroleum additives in the formulations.

PLANT ACTIVITY

Depending on the chemical molecule used in the product, herbicides affect plants in different ways. These different modes of activity are not always apparent on the outside of the plant, but they have a major influence on the success (or failure) of a particular chemical and the ability to mix different products for greater efficacy. The following eight modes of action describe the different ways in which forest herbicides can affect (and control) plants.

Examples of chemicals used in forestry work are given for each category:

- *Cell Membrane Disrupter*: Oxyfluorfen, Paraquat.
- *Respiration Inhibitor*: MSMA.
- *Photosynthesis Inhibitor*: Hexazinone, Simazine, Atrazine.
- *Growth Inhibitor*: Pendimethalin.
- *Lipid Biosynthesis Inhibitor*: Fluazifopbutyl, Sethoxydim.
- *Growth Regulator*: Dicamba; 2,4-D and 2,4-DP; Picloram; Triclopyr.
- *Amino Acid Synthesis Inhibitor*: Glyphosate, Imazapyr, Metsulfuron methyl, Sulfometuron methyl.
- *Pigment Inhibitor*: No forestry chemicals.

FACTORS AFFECTING CONTROL

Plants vary in their susceptibility to different herbicides. They absorb various compounds differently and have different abilities to detoxify the herbicide. Herbicides start breaking down at varying rates soon after application. This breakdown is caused by microorganisms, sunlight and chemical reactions. Herbicides eventually lose all effectiveness.

PESTICIDE APPLICATION

The type of application and equipment to be used for a specific job will depend on a number of things.

Before making a pesticide application you should:

- Know the pest to be controlled;
- Be familiar with pesticides available for use;
- Determine if a Certified Applicator is required;
- Know the size of the area needing treatment;
- Have accessibility to the area;
- Identify the presence of sensitive areas (*e.g.* wetlands, streams, houses, etc.) and organisms (such as livestock, wildlife and any threatened and endangered species);
- Determine the appropriate application method;
- Properly set up and calibrate the equipment to apply materials;
- Apply pesticides only under appropriate environmental conditions; and
- Always read and follow label instructions.

ENVIRONMENTAL CONCERNS

The forest manager must be acutely aware of the risks and consequences of pesticide use and their application in and around forested environments. Use pesticides only when necessary to minimize pesticide impact in areas receiving direct application as well as non-target habitats and organisms which are potential recipients of pesticide drift and run-off.

Before you apply a pesticide, consider these points:

- Do not apply a pesticide in windy or rainy conditions when the chances of drift and wash off/run-off are high.
- Choose an application method and a pesticide formulation that will minimize the potential for movement of the material to off-site locations.
- Restrict or minimize the use of volatile pesticides on areas in or around sensitive on-target plants or animals, especially during hot weather.
- Generally, liquid pesticides applied by broadcast methods are more subject to drift than are granular formulations and their application methods.
- During liquid application, spray droplet size should be maintained within the recommended range for the proposed target and the

application method to be used. In general, large spray droplet sizes (> 300 microns) reduce the potential for pesticide drift. Large droplets do not evaporate as quickly as smaller droplets, so more material will potentially be available to hit the target site, especially during hot, dry weather. However, target spray coverage is usually improved as droplet size decreases (up to a point) since there are many more small droplets than large droplets per given volume of spray material. Another drawback with large droplets is that they may bounce off of and not adhere to leaf surfaces, resulting in poor coverage and increased off-site contamination.

- Use additives to minimize drift and enhance efficacy as appropriate.
- Materials applied to the soil surface can be moved off-site through run-off.
- Individual stem application of pesticides can reduce the possibility of non-target impacts of the pesticide.

APPLICATION TERMINOLOGY

Terms commonly referred to when dealing with methods of applying pesticides in forestry include:

- *Application Rate*: The specific amount of pesticide applied to a treated acre or target system.
- *Broadcast*: Uniform application to an entire area.
- *Banded:* Application to a strip or band over or along each tree row.
- *Basal*: Application to the lower portion of stems or trunks.
- *Cut Surface*: Application to a cut or incision in a tree or to a stump.
- *Directed*: Aiming the pesticide at a specific portion of a plant.
- *Foliar:* Application to the leaves of plants.
- *Over-the-top:* Application over the top of the crop trees.
- *Soil Application:* Application to the soil rather than to vegetation.
- *Soil Incorporation*: Application to the soil followed by tillage to mix the herbicide with the soil.
- *Soil-spot Treatment:* Application to a small area of the soil surface.
- *Stem Injection:* Application into incisions around a tree stem.
- *Stump Treatment:* Application to the top or edges of a tree stump.

Some of the terms that describe the purpose or timing of pesticide applications in forestry include:

- *Cut Surface*: Includes trunk injection, frill, frill-girdle, girdle and cut stump treatment.
- *Desiccation:* The “brown-out” or drying of vegetation by use of herbicides to aid in burning for site preparation.
- *Dormant Spray:* Application before buds open in the spring or after trees are dormant in the fall.

- *Early Foliage Spray:* Application early in the year, but at or soon after full leaf development.
- *Fall Foliage Spray*: Application in late summer to early fall, generally used with readily translocated herbicides.
- *Plantation Weed Control:* Using herbicides for herbaceous weed control to ensure survival and rapid growth of planted tree seedlings.
- *Postemergent:* Used after the crop trees or weeds begin to grow (emerged).
- *Pre-emergent*: Applied before seedlings or weeds begin to grow (emerge) in the spring. This most often refers to applying a herbicide after the trees are planted, but before the weeds begin to grow.
- *Preplant:* Applied before the crop trees are planted.
- *Reforestation:* The process of establishing tree seedlings.
- *Release*: The removal of woody or herbaceous weed competition from developing young stands to improve their growth.
- *Site Preparation:* Preparing an area for reforestation by clearing or other vegetation control.
- *Summer Foliage Spray:* Application to mature foliage later in the season.
- *Timber Stand Improvement:* Selective removal of undesirable trees to improve growing conditions for the desirable residual trees.

Since the primary pesticides used in forested environments are herbicides, the following sections will deal primarily with those materials. However, the application methodology and the calibration of equipment appropriate to apply insecticides and fungicides will involve the same basic techniques but will require different nozzle types, pressures and rates. If one of these other pesticides will be applied, use the procedures listed below and modify according to directions on the pesticide label.

APPLICATION METHODS

Foliar and soil-active materials are often broadcast over the entire area to be treated. They can be applied to the foliage or soil by either aerial or ground mechanical equipment. Broadcast applications are common for site preparation. In some areas this method is used for herbaceous weed control and woody release.

FOLIAR

Many forestry herbicides enter the plant through the green foliage and young stems. Plants that are shielded from foliar sprays by taller or adjacent plants will not be controlled as well as those fully exposed. Adjuvants are added to the spray mixture to aid in the coverage effectiveness or safety of these herbicides. However, some formulations already include adjuvants. Always follow label directions. Adjuvants may be particularly useful for late-season use as foliage becomes waxy and difficult to penetrate.

SOIL

Soil-active herbicides may be applied to the soil as liquid or granular formulations. Control will not occur until there is adequate rainfall or sufficient soil moisture. After rainfall dissolves and moves the herbicide into the soil, it is taken up by the roots of established plants. Pre-emergence herbicides applied to the soil kill vegetation as seeds germinate or new plants grow through the treated ground. Season of the year, soil moisture, texture and pH, as well as organic matter and rainfall, greatly affect soil-active materials.

AERIAL APPLICATION

Aerial application is commonly used to apply pesticides in forestry. This is because tract size is often large, access is difficult; and the vegetation is often tall and dense. Large acreage can be treated more economically and in less time by air. Untreated buffers are established around the perimeter of the treatment area. Firebreaks, flagging, or Global Positioning Systems (GPS) are used to mark treatment boundaries and flight lines. Both fixed- wing aircraft and helicopters are used for forestry applications on insecticides and biological control agents. Forest herbicides are labelled for helicopter application and not by fixed-wing aircraft. Since most states require a separate training and testing for aerial certification, only a brief discussion will follow here.

The use of control droplet aerial (CDA) spray equipment and orienting the nozzles with the air flow causes large droplets. Boom length should be 75 per cent of the total wing or blade span. Nozzles located on booms longer than this can cause excessive drift to occur. The larger the droplet, the less chance of drift to non-target sites. Drift-reducing agents and invert emulsions that change the physical composition of spray mixtures can be used to reduce chemical drift. However, when they are large, droplets may reduce the effectiveness of foliar-absorbed herbicides. Larger droplet sizes display a tendency to bounce off of leaf surfaces. Large droplets tie-up much more spray volume per drop than do smaller droplets. This may cause inadequate coverage, unless greater volumes per acre are applied.

Spray carrier volume should be adjusted to insure effective coverage of target vegetation. Water-based formulations require 5 to 20 gallons per acre (GPA) with the higher carrier volumes necessary when treating multi-story canopies and dense vegetation. Oil emulsions use carrier volumes of 5 to 10 GPA due to costs and deposition efficiency. Aerial applications for midstory and understory hardwood control requires 15 to 20 GPA to insure good coverage beneath closed pine canopies (Minogue 1996). Solid formulations of soil active materials are also applied aerially. Uniform distribution of solid materials is more difficult than with liquid formulations. Finc particles and dust from the granules can increase the risk of off-site drift. To minimize streaks or skips in the treatment area and off-site movement, apply solid formulations only when wind speeds are less than 5 miles per hour.

MECHANICAL GROUND APPLICATION

Ground application equipment can be more versatile than aircraft. They can treat small or large areas, do banded or broadcast application, and are not so limited by weather.

Crawlers, skidders, 4-wheel drive farm tractors and the sturdier ATV's (all-terrain-vehicles) can apply herbicides. The selection depends on the job to be done and the site conditions. Ground machine application has definite limits of terrain and stand conditions.

The pesticide application equipment mounted on the machine must be suitable to do the job. Broadcast type sprayers are most commonly used. The application equipment must be able to cover a sizable area efficiently and must be durable. Each component of a properly working sprayer is important for efficient and effective application. The main limit of ground equipment is usually the presence of brush tall enough to mask a major portion of the spray pattern. Spray coverage of plants must be nearly complete, not just on one side, for effective kill.

For boom type sprayers, flat fan-type nozzles should be used to apply broadcast herbicides. Flat fan nozzles produce an elliptical pattern, where the edges are light and the centre is heavy. These should be spaced on the boom for 30 - 40 per cent overlap. When it becomes necessary to apply herbicides in bands, use an even fan or flood nozzle. These nozzles produce a uniform pattern across the area sprayed. The fan nozzles should be operated at pressures of 20 - 40 pounds per square inch (psi). Flood nozzles are designed to operate at lower pressures 5 - 15 psi. The capacity of both type nozzles should be 15 - 20 gallons per acre (GPA) when operated at 2.2 - 4 miles per hour.

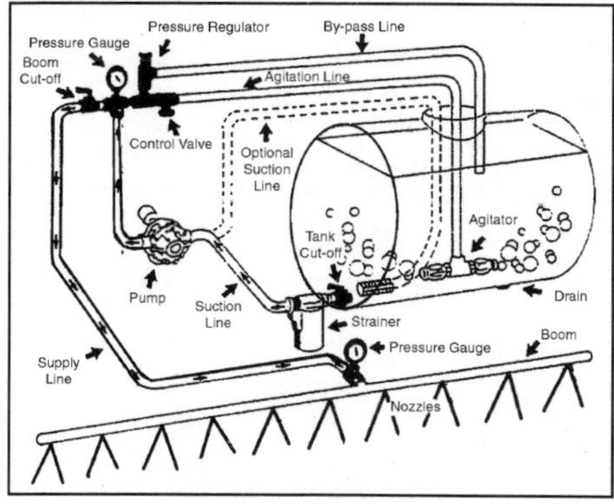

Fig. Typical Boom Sprayer.

An alternative to broadcast foliar application would be to broadcast a soil-active herbicide. This may be in a granular form that can be applied before full leaf growth masks the distribution. Several liquid formulations also have soil activity.

Banded applications are made with herbicides labelled for herbaceous weed control. Some are labelled for application over-the-top of newly planted trees. These foliar or soil-active materials are applied in four- to six -foot wide bands. For resistant perennial species, make a late summer herbicide application at higher rates before seedlings are planted or select a different herbicide.

MANUALLY APPLIED GROUND APPLICATION

Manual applications are usually applied using backpack sprayers, mist blowers, hand-cranked broadcast spreaders, spotguns or one of various injection devices.

The commonly applied manual treatments used in forestry are:

- Directed foliar sprays.
- Basal sprays and stump treatments.
- Tree injections.
- Soil spots and granule/pellet applications.

Directed foliar sprays are best used to release 1- and 2-year-old pine stands when brush competition is less than 6 feet tall. Apply the pesticide spray on the target foliage. Direct the spray away from pine foliage and growing tips. The benefits of release can be lost when herbicides are misapplied to needles and shoots, and pines are damaged. Directed foliar sprays are usually applied with a backpack sprayer and a spray wand equipped with a full cone, flat fan, or adjustable cone spray tip.

Full basal sprays require that the lower 12 to 20 inches of target hardwood stems be completely wet on all sides with the spray mixture. Full basal sprays are effective on target stems. A backpack sprayer is used with a wand or spray gun fitted with a narrow-angle flat fan, cone or adjustable tip.

Streamline basal sprays can control many woody plants including hardwoods up to 2 inches in diameter at breast height (dbh). Trees of susceptible species up to 6 inches in diameter can be controlled. However, treatment of small hardwoods less than 2 inches dbh results in the most control.

For stems less than 2 inches dbh, apply the stream of spray up and down single stems for about 6 to 8 inches, or spray across multiple stems creating a 2 to 3-inch-wide band. Direct the spray stream to smooth juvenile bark at a point about 6 to 24 inches from the ground. Stems that are beyond the juvenile stage, thick barked, or near 3 inches in diameter require treatment on both sides, unless they are susceptible species. Back-and-forth bands can also be sprayed on larger stems. Apply in late winter and early spring when leaves do not hinder spraying the stem. The best application time will depend on the herbicide, species and location. Avoid applications in young pine plantations on hot days if an ester herbicide formulation is used because pine injury may occur from vapour drift.

Tree injection can be used alone or in combination with other individual stem treatments for site preparation, pine and hardwood release, timber stand improvement, stand conversion and creating cavity trees for nesting. This physically-demanding method requires workers who can repeatedly and precisely chop into tree trunks deep enough to properly deliver herbicide for uptake in the sap flow. Frequent sharpening and maintenance of injection tools is needed for best results.

Commonly used tree injection methods are:

- The hack-and-squirt,
- Hypo-hatchets, and
- Tubular tree injectors.

HACK-AND-SQUIRT

The hack-and-squirt is an effective and economical means of selectively controlling undesirable hardwood stems. A lightweight hatchet is used to cut into the tree stem through the cambium and a herbicide is sprayed on the cut from a trigger squeeze bottle.

HYPO-HATCHET

The hypo-hatchet is a hatchet with an internal herbicide delivery system connected by a hose to an external herbicide container. When the hatchet strikes a tree, the blade must penetrate into the sapwood. The impact of the striking action drives a piston forward that delivers 1 ml of herbicide into the cut. The rate cannot be adjusted. Daily cleaning and lubrication of the impact piston is required maintenance, along with periodic replacement of rubber O-rings and seals. Always wear safety glasses when using the hypo-hatchet because of frequent herbicide splashes. CAUTION: All hoses and fittings should be checked daily for leaks and appropriate repairs made to prevent applicator exposure.

TUBULAR TREE INJECTORS

Tubular tree injectors have a long metal tube fitted with a chisel-type blade that is used to cut through the tree bark into the sapwood near the base of the tree. The unit is equipped with a lever, handle or wire, which is pulled to deliver the herbicide (usually 1 ml) from the cylinder into the cut. The delivery rate can be adjusted for accurate calibration.

Waist-high injections by the hypo-hatchet and hack-and-squirt methods are just as effective and as fast to perform as basal injections. With larger stems, apply more herbicide by basal injections because of the larger groundline diameter compared to diameter at breast height.

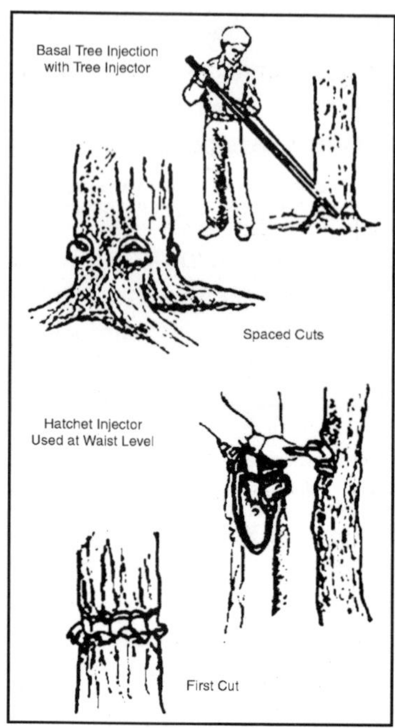

Fig. Tree Injector Equipment.

TREATING STUMPS

Treating stumps with herbicide can prevent resprouting of many species. This can be an effective, lowcost treatment following harvest for site preparation and after partial cuts for timber stand improvement. Hand clearing treatments using saws or axes for pine release can be enhanced by treating the stumps with herbicide to prevent regrowth.

A backpack sprayer can be used that has a wand or spray gun equipped with a straight stream, fan or hollow-cone nozzle. Alternatively, a sawyer can carry herbicide in a utility spray bottle for treating stumps after cutting; or use a wick applicator for small-diameter stumps.

Treat freshly cut stumps as soon as possible after cutting. For stumps over 3 inches in diameter, completely wet the outer edge, or cambial area, with the herbicide. Smaller stumps are usually completely wetted. To be successful, treat all small stumps. The sawyer or companion applicator should treat soon after felling so no stumps are skipped. Treat older, cut stumps with the streamline mixture. The mixture is applied to the outer 1-inch edge of the stump until run-off and to the base of any sprouts. Stump treatments within four hours of cutting have been shown effective—the sooner the better. Spots of soil-active herbicide are applied to the soil surface in grid patterns or around target stems for site preparation and pine release. This method is effective in controlling stems up to 10 inches dbh. Apply exact amounts of herbicide, specified in milliliters (ml), to the soil surface at prescribed spacings. The effectiveness of the treatment depends on the applicator's accuracy and consistency in amount applied and spacing.

Spots are applied to the soil by using a spot-gun or a spray-gun equipped with a straight-stream spray tip. The spotgun delivers a set amount while the spray-gun method requires training to judge the amount applied. A spotgun is an adjustable graduated cylinder or syringe operated by squeezing the handle. A forceful squeeze can project spots up to 15 feet. A spray gun uses pressure from the backpack sprayer to project spots to over 20 feet, requiring less exertion. Both can be connected to a backpack sprayer, and the spotgun can also be connected to a side-pack container.

GRANULES AND PELLETS

Granules and pellets can be applied by hand-cranked spreaders, air-blown backpack spreaders, and hand-broadcast.

Hand-cranked broadcast spreaders can distribute granular or pelletized herbicides on small tracts and areas with steep slopes or rough terrain. They can be used where machine spreaders are not suitable. Advantages of hand-operated spreaders are that they are small, simple, inexpensive and generally reliable hand tools. Unfortunately, uniform application is often difficult to obtain, and treatment is slow and laborious.

OTHER CONSIDERATIONS

The forest manager must be acutely aware of all of the environmental and personal safety concerns associated with using pesticides. Federal and state efforts to protect individuals, wildlife and the environment from harm and contamination are becoming important issues to determine which pesticides will be registered and for what use.

The forest manager must be aware of both current and developing limits and restrictions dealing with pesticide use, and must use and enforce all safety precautions and environmental safeguards. Pesticides that are incorrectly released into the environment (whether during application, mixing, loading, equipment cleaning, storage, transportation or disposal of pesticides) pose a threat to individuals, wildlife, endangered species and both surface and groundwater. Forestry pesticides are often applied to large areas which frequently consist of diverse habitats encompassing streams, rivers, estuaries, swamps or open water. These diverse habitats may be home to humans; domesticated animals; and terrestrial, aquatic and/or marine organisms. Consequently, special limits and restrictions often apply to pesticide use in forests. Always read and follow label directions.

Report fish or wildlife kills in pesticide-treated or adjacent areas to the appropriate Natural Resource agency. It may want to investigate the reason for such a kill to help prevent future occurrences. Many conditions other than pesticides can kill fish or wildlife.

BENEFICIAL FOREST INSECTS

There are many species of beneficial forest insects. Some of these insects feed on forest debris and aid in its deterioration; others feed on organic matter in the duff and soil and contribute to improvements in soil fertility. Many others are parasites or predators of destructive insect species. Many insects are important food sources for birds and other small animals. Bees and certain other insects are important pollinators of many commercial crops as well as forest plants and trees.

Forest managers and pesticide applicators must be aware of these and other wildlife members of the forest environment. Pesticide labelling gives useful information about toxicity to non-target life forms. Learn as much as possible about the health and environmental hazards of the pesticides that may be used. Select the pesticide and application method that will have the least adverse impact and still get the job done.

PRESCRIBED BURNING

This is often employed to help control undesirable forest species. It also can help reduce brown spot needle blight on longleaf pine seedlings and annosum root disease as well as other undesirable forest conditions. Prescribed burning is often an important part of most chemical site preparation treatments. Not only does it add to the herbicide kill, it helps clear the site to facilitate reforestation work. When properly timed and executed, prescribed fire has little adverse effect on the environment. However, prescribed burning and other non-chemical pest control measures can present risks and have undesirable effects. Prepare a written, prescribed burning plan before each burn to identify measurable objectives for burning and specific conditions under which the burn will be conducted. Be sure to make a smoke management screening evaluation and conduct a follow-up evaluation of the effectiveness of your prescribed burn. In most states, you must contact the local Forestry Commission/Department office for a burning permit before you start the burn.

ENDANGERED SPECIES ACT

The Endangered Species Act provides:

- Legal protection for endangered and threatened species.
- Requires all Federal Agencies (for example, EPA) to ensure their actions will not jeopardize the existence of any endangered species.

About 58 endangered species, or 17 per cent of the total endangered species currently listed, occur in forest situations in the United States. Many of the pesticides presently labelled for use in forests are considered to have an adverse affect on one or more of these endangered species. These numbers doubtlessly will change over time, but they indicate there are many endangered species found in forest situations, and a number of pesticides will affect them.

Since 1988, every affected pesticide has a warning on the label:

- Label prohibits pesticide use in occupied habitat of endangered species.
- Ranges are identified to county level.
- They are to be used in identified counties permitted only if not within the range of endangered species.

An information bulletin should be available in those counties of a state listed on the label. The Information Bulletin will have a county map giving the boundaries of those areas of the county where the use of the pesticide will have some restrictions, the endangered species affected, and a description of its habitat. Information bulletins should be available through local county Extension and state forestry offices. Each state will have an "enforcement plan" to implement the Endangered Species Act.

7

Forestry Conservation Management

Forestry Conservation Management Practices (CMPs) are specific, science-based guidelines for conservation of rare species during forest harvesting. CMPs are somewhat analogous to Forestry Best Management Practices (BMPs), except whereas BMPs focus mainly on protection of water resources, CMPs specialize in protection of rare wildlife. The primary objective of CMPs is to guide harvesting activities such that rare species listed under the Massachusetts Endangered Species Act (MESA) are not impacted in a way that jeopardizes long-term viability of local populations. CMPs first identify and describe potential impacts of forest harvesting to state-listed species, whether impacts may be direct (*e.g.*, physical injury or death of individual animals) or indirect (*e.g.*, alteration of habitat in a way that reduces overall reproductive success of a local population).

Then, CMPs provide specific guidelines to avoid or minimize impacts that would be considered negative or potentially detrimental to a local population; the guidelines are based on scientific knowledge of the habitat requirements, reproductive strategy, dispersal ability, survivorship, and other ecological factors that influence population dynamics of the species. CMPs are based on the principle that protection and enhancement of the long-term viability of local populations is prerequisite to statewide recovery of imperiled wildlife.

Simultaneous with the objective of protecting local populations of state-listed species, CMPs aim to maintain adequate opportunity for sustainable management of timber products in Massachusetts. To this end, CMPs tend to focus forest harvesting restrictions on the critical areas within known habitat of state-listed species, thereby allowing timber management to proceed with fewer restrictions over as large an area as possible.

This strategy is based, in part, on recognition that forest harvesting typically results in temporary habitat change or sometimes even habitat improvement rather than permanent habitat loss. Thus, the CMP strategy is designed to help maximize the protection of state-listed species and the ability of Massachusetts landowners to manage their forests for timber and other wood products.

GENETIC RESOURCES IN CONSERVING FOREST

A wide range of methods, from protected reserves to intensive management of breeding populations for production systems, can be used to conserve FGR. The choice of methods depends on available genetic material, selected time scale and specified aims. The method selected and the subsequent implementation of a conservation strategy also depend on the availability of both human and financial resources. The two most commonly considered methods for conserving FGR are *in situ* and *ex situ* conservation. The term *in situ* refers to the continued maintenance of tree populations at their natural sites, in the environment to which they have adapted. *Ex situ* conservation takes place outside the natural habitat of a tree species, and may consist of activities such as establishing live collections or *ex situ* conservation stands, or storing seeds, pollen or tissue.

FGR can also be conserved and maintained by using tree species in forestry or other land-use systems such as agroforestry. It is likely, however, that forest management interventions reduce genetic variation in tree populations. Not even carefully planned and implemented forest management activities, therefore, can replace more active conservation measures with clearly specified objectives. The same caveat also applies to tree domestication and the use of trees in agroforestry systems.

IN SITU CONSERVATION

There is a general consensus among scientists and practitioners that no single conservation method is adequate, and that different methods should be applied in a complementary manner. *In situ* conservation, however, has a number of benefits and so often forms the basis of conservation programmes. It allows evolutionary processes to be maintained, including the adaptation of tree populations to changing environmental conditions. This is particularly important for breeding programmes, since future human needs and environmental conditions are difficult to predict. The reproductive biology and overall survival of many tropical forest species are dependent on complex ecological interactions. Thus long-term genetic conservation of such species is difficult, if not impossible, without *in situ* conservation. In addition, recalcitrant seed behaviour (*i.e.* inability to tolerate desiccation) also complicates long-term *ex situ* conservation efforts for many tropical tree species. *In situ*conservation can also contribute to the conservation of biological diversity at higher levels, *i.e.* species and ecosystems.

Protected areas have often been established on the basis of ecosystem or species conservation, rather than gene conservation. Thus the design of *in situ* conservation programmes has been considered primitive. In tropical forests in particular, the complexity of interactions and lack of scientific information have hampered the development of *in situ* conservation strategies. Uncertainty surrounds the adequate size of *in situ* conservation areas, the number of individuals to be included, how to select locations, and how genetic variation is distributed within selected areas. Undisturbed tropical forest ecosystems are often taken as a starting point when planning *in situ* conservation programmes. Today, however, rural landscapes are a mosaic of disturbed and less disturbed patches of forest, ranging from seemingly natural and secondary forests to seriously degraded forests and other wooded fragments.

The most obvious genetic effects of fragmentation are a loss of genetic diversity at both population and species levels, changes in the genetic structure of a population and increased inbreeding. Clearly, *in situ* efforts cannot always rely on intact natural forests, and so it is essential to understand how the processes of genetic drift, gene flow, selection and mating affect genetic diversity in fragmented forests.

IPGRI's partners at the Universities of Costa Rica, Alberta and Massachusetts have been studying the effects of fragmentation on the genetic diversity of *Enterolobium cyclocarpum* in Costa Rica. Studies of the reproductive biology of *E. cyclocarpum*, a tropical dry forest species, in continuous forests and forest fragments in pastures indicate that pasture trees are less likely to receive pollen and set fruit, and that their fruits bear less seed, than trees located in continuous forests. Also, outcrossing rates within the two groups of trees have been found to be similar, but progeny vigour among seedlings in continuous forest is higher than in pastures. These results have immediate implications for the conservation and management of *E. cyclocarpum* and other species in similar habitats. Trees in pastures can aid the movement of pollinators between intact forest fragments and so contribute to gene flow and maintenance of genetic diversity among forest fragments. However, because seedlings in pastures have less vigour, seeds from these areas cannot be recommended for establishing plantations or rehabilitating degraded natural forests.

In southern India, sandal (*Santalum album*) forests have long been exposed to selective logging, poaching and changes in land use. The effects of disturbance on the genetic diversity of sandal have been investigated by the University of Boston, together with the Ashoka Trust for Research in Ecology and the University of Agricultural Sciences in Bangalore. Results from two protected areas indicate that genetic diversity of sandal is highest in the core undisturbed park zone, whereas allelic diversity is reduced in the outer buffer and disturbed zones. These findings suggest that present protection measures are adequate and that excessive logging will cause genetic deterioration in natural sandal populations. Allelic diversity, however, is similar in the two disturbed zones, which suggests that, contrary to general assumptions, no significant genetic segregation has resulted from different degrees of disturbance. This finding can be attributed to the fact that sandal is insect-pollinated and animal-dispersed, *i.e.* substantial gene flow still occurs despite varying degrees of disturbance.

EX SITU CONSERVATION

In situ and *ex situ* conservation should be used in a complementary fashion to conserve FGR. Both are an integral part of the conservation process, and both can only be effective after genetic diversity has been located and conservation priorities have been set. The main purpose of *ex situ* conservation is to capture and maintain a representative sample of the existing genetic diversity of a species. For highly endangered tree species, *ex situ* conservation may be the only approach in the short to medium term. The main pitfalls of collecting germplasm samples for *ex situ* conservation are: i) limited coverage of genetic variation; ii) biases in the collected plant material; and iii) samples that are too large to deal with. Because *ex situ* conservation is more costly than *in situ*conservation, it is particularly important that sampling of populations and germplasm within populations is given special attention to maximize the use of limited financial and human resources.

The traditional approach to *ex situ* conservation of FGR is to establish conservation stands of a species outside its native habitat to facilitate gene management. However, long-term maintenance of the collected genetic variation in *ex situ* stands tends to be complicated by genetic drift and potential contamination by external gene flow. Another commonly used method of *ex situ* conservation is to store seeds collected from a range of natural populations. In the case of many tropical tree species, however, this method is limited by recalcitrant seed behaviour. More than 70 per cent of commercially valuable tropical tree species are estimated to have recalcitrant or intermediate seeds. In recent years, therefore, considerable effort has been expended in developing *in vitro* techniques for *ex situ* conservation of recalcitrant tree species. A notable example is cryopreservation. Cryopreservation, the storage of cells or tissue at ultra-low temperatures, offers great opportunities for longer-term *ex situ* gene preservation. Its applicability, however, has been limited by difficulties in identifying protocols to reduce the water content of tissue before freezing. Tissue water content must be reduced to avoid lethal ice formation in cells. Different results from different laboratories engaged in drying the same species of seed are an additional obstacle. Although it is likely that further research will ease these problems, the main application of cryopreservation seems to be in supporting tree improvement programmes and conserving biotechnically derived germplasm, rather than as a widely applied *ex situ* conservation method to support *in situ* conservation of FGR.

Micropropagation, the use of tissue and organ cultures for organogenesis and somatic embryogenesis, can also be applied to maintain genetic variation in germplasm collected for *ex situ* conservation. Practical propagation applications are already available for several tropical trees, based on shoot tissue culture in broadleaved tree species and on somatic embryogenesis in conifers. Micropropagation does not necessarily require expensive laboratory facilities, and can therefore be used in less-developed conditions. Maintaining juvenile material, however, can be problematic. At present, the technology used in somatic embryogenesis is in most cases too expensive for cost-effective *ex situ* conservation.

Since 1997, the Danida Forest Seed Centre (DFSC) and IPGRI have cooperated in enhancing *ex situ* conservation methods for recalcitrant tropical forest trees. Project activities have focused on determining whether seeds are recalcitrant, intermediate or orthodox, and have included seed development research to assess optimal conditions for seed collecting, germination and storage. An international Forest Tree Seed Research Network has been established under the project to facilitate information exchange among scientists developing protocols for collecting, handling, testing and screening of tolerance to desiccation and optimal storage conditions. At present more than 20 countries worldwide are involved in the project. More than 50 tropical forest species have been screened so far. Activities also include publications and training workshops to increase research capacity, particularly in developing countries.

MANAGEMENT OF FORESTLAND IN MASSACHUSETTS

Management of forestland in Massachusetts involves multiple stakeholders with multiple objectives, so communication among stakeholders and development of management standards are essential to guiding compatible and sustainable uses of the land. State-owned forestland in Massachusetts received Forest Stewardship Council certification in May 2004, and this certification requires that there be coordination of forest management among the three state forest management agencies (Division of Water Supply Protection [DWSP], Bureau of Forestry [BOF], and Division of Fisheries and Wildlife [DFW]). In addition, there is a need for state forestland managers to work closely with the Natural Heritage and Endangered Species Programme (NHESP) to develop and implement a formal and consistent process for determining appropriate mitigation of potentially adverse impacts of forest harvesting to state-listed species on both state and privately-owned forestlands.

In light of these goals, CMPs are needed to ensure adequate protection of state-listed species during forest-harvesting operations. Without CMPs, a landowner and/or forester must develop a Forest Cutting Plan without having prior knowledge of possible harvest restrictions required by the NHESP to mitigate potential impacts of forest harvesting to state-listed species.

Under these circumstances, activities proposed in the Forest Cutting Plan must often be modified after the plan is reviewed by the NHESP, which tends to result in delay of the harvest, changes in the amount and distribution of timber to be harvested, or changes in the number, type, or location of stream and wetland crossings. However, CMPs shall make potential harvest restrictions available to the landowner and forester before a Forest Cutting Plan is developed. This gives the landowner and forester an opportunity to develop a plan that is more likely to be approved by the NHESP without need for modification, thereby saving the landowner time and money. With proper planning and attention to detail, use of CMPs shall maximize the efficiency of the cutting-plan filing and review process and enable landowners to better understand how forestry and conservation of rare wildlife can be integrated as compatible uses of Massachusetts forestland.

Benefits of CMPs

CMPs benefit both state-listed species and forest management by (1) being based on scientific knowledge of the ecological factors influencing population dynamics of state-listed species, thereby enabling focused protection of key habitats and other resources, (2) improving the predictability and consistency of review of Forest Cutting Plans by the NHESP, and (3) providing detailed guidance to landowners and foresters for developing Forest Cutting Plans that have high probability of being approved by the NHESP without modification. An additional benefit, which is a byproduct of the first listed above, is that time-of-year harvesting restrictions mandated by CMPs tend to apply to smaller areas than those to which restrictions applied in the past. In this sense, intensive conservation measures are concentrated, which should be more effective in achieving conservation objectives than a strategy by which lighter measures are applied over a broader area. Furthermore, this strategy affords the landowner greater liberty in managing a larger proportion of a given tract of forest.

Process for Developing CMPs

The NHESP is the primary partner responsible for development of CMPs. However, staff of the Natural Resources and Environmental Conservation (NREC) Extension Programme at the University of Massachusetts Amherst and foresters of the DWSP, BOF, and DFW collaborate with the NHESP in this effort.

This group of scientists and land managers meets regularly to work together on every aspect of the CMP development process. The first step in developing CMPs is to determine which state-listed species should be addressed. This decision is based largely upon the frequency with which forest harvests are planned in habitat of each state-listed species, whereupon species that most frequently coincide with harvests are considered top candidates for development of CMPs.

A second factor is the degree to which protection of each state-listed species is expected to impact the logistics of forest harvesting, whereupon species that are likely to require highly protective measures are considered top candidates. A third factor, which often is related to the second, is the sensitivity of a state-listed species or its habitat to forest harvesting, whereupon species that are highly sensitive are considered top candidates. When multiple species have similar ecological requirements and conservation strategies, groups of species may be lumped into a single CMP document (*e.g.*, Blue-spotted Salamander, Jefferson Salamander, and Marbled Salamander are lumped together in a "Mole Salamander" CMP document).

When there is agreement between the NHESP and its collaborators about which state-listed species to address in a CMP document, the NHESP conducts a thorough search of scientific literature pertaining to the ecology of the species of interest. The information is then consolidated into a species account that includes a physical description of the species and a discussion of its range, distribution, habitat requirements, seasonal movements, reproductive strategy, survivorship, and other concepts associated with its life history. Additionally, the account includes a discussion of known and potential threats to population viability, including those posed by forest harvesting.

The NHESP and its collaborators then develop the actual management practices that scientific evidence suggests is necessary for successful conservation of the species. These management practices are specific to forestry-related activities and, like Forestry BMPs, are split into two categories: required management practices and recommended management practices. Required management practices (denoted with "R" in the CMP document) are those which should be considered mandatory during most harvests, barring special circumstances. Recommended management practices (denoted with "G" in the CMP document) are optional guidelines that are expected to provide additional benefits to the species of interest.

DEFINITION OF TREES OUTSIDE THE FOREST

Trees outside the forest are defined by default, as all trees excluded from the definition of forest and other wooded lands. Trees outside the forest are located on "other lands", mostly on farmlands and built-up areas, both in rural and urban areas. A large number of TOF consist of planted or domesticated trees. TOF include trees in agroforestry systems, orchards and small woodlots. They may grow in meadows, pastoral areas and on farms, or along rivers, canals and roadsides, or in towns, gardens and parks. Some of the land use systems include alley cropping and shifting cultivation, permanent tree cover crops (*e.g.* coffee, cocoa), windbreaks, hedgerows, home gardens and fruit-tree plantations.

Classification of trees outside the forest presents certain difficulties. There are existing classifications for agroforestry, but none applicable to all trees outside the forest. For practical reasons, the FRA 2000 definition of "forest" combines aspects of both land cover and land use. This approach creates difficulties not only for classification of forest, but also for classification of TOF.

In a study on data-gathering on TOF in Latin America, where classification was primarily based on land use criteria, separating land use and land cover aspects was found to be a main source of misinterpretation. There was a possibility of confounding coffee plantations and trees in pasture with forest, given their high density.

This clearly shows some of the problems involved in establishing a simple and reliable *a posteriori* classification.In France, the National Forest Inventory (IFN) and Teruti Land Use Study-have begun to attempt coordinating classifications of trees outside the forest. The objective is eventually to use the annual Teruti data to update the IFN ten-year data, with a single national nomenclature as a possible end result.

FUNCTIONS AND CHALLENGES

In industrialized countries, farmers list shade and shelter, soil protection and improvement of the landscape and rural environment as their main reasons for growing trees. In the tropics, farmers grow woody species for food security and subsistence. Trees outside the forest are a major source of food. Livestock fodder produced by TOF can be a matter of life and death in semi-arid or mountainous areas.

Fuelwood remains the prime source of energy in developing countries, representing up to 81 per cent of the wood harvest (FAO 1999). In contrast, in the industrialized countries, fuelwood accounts for less than 10 per cent of total fuel consumption (FAO 1998). Very few studies have reported on overall fuelwood output from stands and single trees outside the forest, but agroforestry systems and orchards are known to provide a large part of the resource.

Trees outside the forest have an important ecological role. Planted trees and shrubs in fields help to check run-off and erosion and control flooding, as well as helping to purify water and protect against wind. Trees lining rivers and streams help to maintain biodiversity, providing spawning beds for fish and shellfish and shade which reduces eutrophication.

The unique role of trees in soil protection and conservation, checking wind and water erosion and maintaining soil fertility is universally acknowledged. Also important are the cumulative benefits of trees on smallholdings to soil and water conservation, in particular in the larger context of mountain watershed management; their positive impact on climate; and their role in buffering the effects of desertification and drought.

RESULTS OF SELECTED STUDIES

In spite of the limits of data at the regional or global level, a number of local initiatives have been carried out. The approaches of the various studies differ in accordance with the purpose and scale of the analysis. Few studies use methods resembling the conventional forest inventory. Many studies rely on existing literature or estimates drawn from surveys and interviews. The quantification of products is often based on different parameters, such as estimates of global output, marketed output, observed or potential productivity or economic value. Thus the reliability of the results is uncertain.

The following are the results of some national initiatives that have assessed trees outside the forest.

In Kerala, the most densely inhabited state of India, a study estimated that of the total annual production of 14.6 million cubic metres of wood in the state, about 83 per cent was from homesteads (house compounds and farmlands), 10 per cent from estates (plantations of rubber, cardamom, coffee and tea) and only about 7 per cent from forest areas 26.6 per cent of the state area is under forest cover. Trees outside the forest met about 90 per cent of the fuelwood requirements of the state. Fuel from coconut trees alone, including both wood and non-wood materials (pruned and fallen), constituted about 70 per cent of the total fuelwood supply.

A study in Haryana State in India, an intensively cultivated state with about 3.8 per cent of its area classified as forest land but only about 2 per cent under actual forest cover showed that farm forestry (trees along farm bonds and in small patches up to 0.1 ha) accounted for 41.2 per cent of the total growing stock of wood. Multiple tree rows along roads and canals accounted for 13 per cent and 9.6 per cent, respectively; village woodlots for 24 per cent; and block plantations of less than 0.1 ha for 10.6 per cent.

In Morocco, where forest cover is less than 5 per cent of the land cover and other wooded lands only 7 per cent, nearly 20 per cent of the land may be occupied by trees outside the forest, namely as wooded pasture (84 per cent) and fruit-tree plantations (12 per cent) (Rosaceae, citrus, olives trees, palm trees, walnut trees, fig trees, almond trees).

Fruit production has an important place in the national economy. It is noteworthy that even when a forest is largely destroyed, the carob is one of the few species traditionally conserved, as it is highly appreciated by farmers for multiple purposes, providing both fodder and income from the sale of its fruit for export. However, there are no reliable data on the distribution and potential of this "forest" resource, which is of interest for farmers, herders, concessionaires and the government and distributed on agricultural and forest lands.

In the Sudan, the National Forest Inventory has undertaken a national land use inventory to provide area and volume statistics for planning at the subnational and national levels. The inventory was designed to provide preliminary estimates regarding products other than the traditional fuelwood and timber, such as the amount of gum, fruit or nuts that can be collected and the distribution of non-wood species of interest. In Costa Rica, the Tropical Agricultural Research and Higher Education Centre (CATIE), in collaboration with Freiburg University, Germany, is developing a regional methodology for Central America to assess tree resources outside the forest.

A mix of satellite remote sensing, aerial photos and ground sampling is used to address the complexity of the resource and to allow dynamic monitoring of resources at the national and regional levels data-gathering on TOF in eight Latin American countries. None of these countries had established a database and the search for information was multisectoral. The statistics on land cover and land use gave some idea of the relative importance of trees outside the forest.

In Kenya, extensive tree planting on farmlands was promoted in the 1970s and 1980s, with land tenure security as a major incentive. There is an increasing trend of tree cover and species diversification on privately owned farms. Assuming that the present rate of increase in tree planting will continue, it was estimated that farms produced about 9.4 million cubic metres of wood in 2000 and will produce about 17.8 million cubic metres in 2020.

Their share of the total wood produced in the medium- and high-potential districts was projected to increase to 80 per cent in 2020. Indeed, while natural stands of trees have declined there has been a corresponding increase in tree planting in much of the densely populated plateaus of Kenya. As natural forests are reduced or become inaccessible, agroforestry systems help people to diversify production and income and to protect themselves from shortages of fuel and wood.

In Bangladesh, natural forest formations cover less than 6 per cent of the country and the population growth rate is extremely high. An inventory of homestead/village forests in the country indicated that trees outside the forest constitute a vital resource for local populations, providing food, fodder and fuelwood. The sampling method was based on dual village/household sampling with an agro-ecological and administrative sampling base. Rural Bangladesh was divided into six major regions considered as agro-ecological strata, each subdivided into *thanas* (administrative entities, subdistricts). The households making up the sampling units were chosen at random from a number of villages.

The inventory sampled data on palm trees and cane as well as trees, bamboo and thickets. The results, expressed per stratum and per inhabitant, provide volumetric data for fuelwood and sawnwood and species data for total amounts under and over 20 cm. This inventory was apparently the first to nationwide assessment of trees outside the classified forests in Bangladesh.

METHODS AND TOOLS FOR FUTURE ASSESSMENTS

A priority challenge of future assessments is to know the state and dynamics of all tree resources both in and outside the forest. A country embarking on a planning exercise cannot confine itself solely to the trees within its forests, especially when its wood resources appear to be insufficient.

The choice of tools and methods used to describe or assess trees outside the forest depends on the scale of analysis, kind of data and degree of exactitude desired. The tools used are not generally specific or new; rather, they are combined and implemented in original ways.

The inventory in Bangladesh described above is one of the numerous examples of methods developed for gathering data on TOF. The Bangladesh study gives evidence of the adaptations needed to bring conventional forest inventory procedures in line with the specific nature of this resource.

In some aspects - *e.g.* structure, spatial distribution and extent of area cover - trees outside the forest are more difficult to assess than forest formations. The assessment of TOF does not lend itself to the potential cost savings associated with expanded uses of remote sensing technology. Remote sensing by satellite presents more difficulties for assessing TOF resources than for assessing attributes such as forest area. However, satellite data do allow a region to be stratified on the basis of ecological criteria and land cover, providing the basis for a good working document for more specific work in the future. The most commonly used remote sensing technology for TOF resources is aerial photography, which can be used to describe spatial distribution and to distinguish TOF cover classifications, providing the appropriate scale is chosen. However, high costs prohibit widespread use of aerial photography for TOF assessments in most countries. The new 1 m resolution satellite sensors represent a possible future alternative to aerial photography.

Some TOF field inventories are modelled on forest inventory methods and keep to biological and physical criteria; others emphasize social aspects, choosing villages as the sampling units. For measurements on the ground, sampling arrangements designed for forest stands may not be the most effective arrangements for trees. Less traditional sampling plans which would theoretically be better suited to this resource should be tested on various categories of TOF, especially those covering fairly large areas.

Studies of the social and economic benefits or impacts of TOF often rely on household surveys, interviews or standardized appraisals such as rapid or participatory rural appraisal. The integration of the last two approaches - biophysical inventory and socio-economic analysis - is not simple and calls for caution given the great variety of social situations that are only meaningful in the local context.

Environmental benefits or impacts of TOF might be indirectly assessed by linking measurable indicators, such as the number and type of trees, with environmental variables such as water quality or erosion. In an urban setting, tree cover might have direct impact on the ambient temperature. Measuring the environmental impact of tree management is an issue for all natural resource planning or management operations. Assessment of trees outside the forest requires geographical, ecological, biophysical, social and economic data.

However, this implies that an important amount of information will have to be carefully processed. The diversity of end-uses for this information, including land use planning and analysis based on inventory, will need to be considered in data assembly and processing and in the presentation of results.

It is important to know the status of trees outside the forest at any given moment, but it is even more essential to be able to trace patterns of change over time in the same area. The two most commonly used approaches have been comparison of aerial photos taken at sufficiently long intervals and surveys among villagers/managers combined with field inventories.

Some countries, such as France and the United Kingdom, have undertaken periodic inventories based on the establishment of permanent plots linked to permanent forest inventories. However, the high cost of this type of operation limits the number of countries able to adopt it. India and Bangladesh are now experimenting with options for the future.

The current trend towards decentralized authority in land use planning suggests the importance of carrying out assessments at the local level, where the geographical, historical and socio-economic context is relatively harmonious. A minimum number of common rules concerning methods and arrangements is necessary, however, if the data are to be comparable at the country level. Certainly, the technical side of assessing trees outside the forest is complex and more research is needed to better pinpoint the resource.

CHALLENGES IN FOREST LAND MANAGEMENT

The Forest Department also faces a great challenge in managing the forest resources of the country. Although 40 per cent of the country's land resources has been protected by legislation and gazetted as forest reserves, huge tract of forested areas (35 per cent of the country's land resource) remains outside of these reserves and have been the subject of different land development. The 1989 National Forest Policy calls for the gazettement of additional 15 per cent of the country's land area under Forest Reserve – making 55 per cent of the country's land area under sustainable forest management. To be specific, the Forestry Department has been continuously facilitating the gazettement of additional land area within the Inter-Riverine Zone to meet the National Forest Policy target and to protect the huge investment made by the Government in timber plantation.

It is an accepted fact that land is a scarce-resource in Brunei Darussalam just like the rest of the world and the modernization/development of other sectors such as agriculture, fisheries, industrial, and housing sector have continuously posed for the conversion of these forested areas into other land-uses.

FOREST CONSERVATION AND PROTECTION INITIATIVES

The Tropical Rainforests are home to 50 per cent of all the animal and plant species on Earth. They are not only providing shelters and habitats for wild fauna and flora but also giving direct benefit to human being by acting as a giant lung for the Earth - absorbing large amounts of carbon dioxide and producing oxygen in exchange.

In addition, the rainforests also provide us with many foods, raw materials, and medicines. Despite their importance, the rainforests, which are known to cover only 6 per cent of the Earth's surface, are always facing threats of being destroyed. Every year about 20 million hectares of rainforest are being cleared worldwide due to logging activities and clearing of land for agricultural purposes. Most of the world's remaining rainforests are in developing nations and the financial return on timber and farming is crucial for survival of their people and thus to a certain extent takes precedence over rainforest conservation. As a result, destruction of the tropical rainforests continues at an alarming rate.

In Brunei Darussalam, the pressure on the natural forest is not that great as compared to other countries within the region, as economically, Brunei Darussalam is still heavily dependent on the production of oil and gas as the main source of income for the country. However, the pristine forest of the country should be conserved and managed in a proper and sound manner, within the context of sustainable forest management, for its social, economic, and environmental conservation values. Over the years, the Forestry Department has been conducting various initiatives, projects and programmes in trying to ensure that our forest resources will continue to be developed and managed in a sustainable manner.

The Heart of Borneo [HoB] Initiative

One of the major on-going initiatives implemented by the Forestry Department is the Heart of Borneo Initiative. The Heart of Borneo initiative aims at conserving a large tract of the forest resources - mainly upland rainforest, spanning the central highlands of Borneo, and extending through the foothills into adjacent lowlands where possible to retain ecological connectivity - involving the voluntary participation of three countries namely Brunei Darussalam, Indonesia and Malaysia. This initiative is considered as one of the largest tropical rainforest conservation initiative in this region. The implementation of HoB initiative in this country actually complements and strengthens the already existing efforts and initiatives of the Department in ensuring its forest resources are managed and developed in a sustainable manner. As part of the country's commitment and support to the implementation of this HoB initiative, Brunei Darussalam has committed to allocate 58 per cent of its land area to be under HoB management. The conservation vision as envisaged by the HoB initiative is actually nothing new. However, with the international recognition and accolade received by this initiative, it had given the already existing conservation efforts implemented by the Forestry Department an added value under the branding of HoB initiative.

The impact of HoB initiative in the implementation of forest management and conservation in the country has been significant. In the past, forest planting activities were organized mainly by the Forestry Department as part of its annual activities, and the private sectors and other relevant stakeholder involvement in the implementation of such activities came in the form of passive reaction as a gesture of support to such planting activities. But now, under the HoB branding, a new dimension in the implementation of forest rehabilitation and replanting activities emerges. The private sectors and relevant stakeholders, now no longer become a passive partners in the implementation of such forest replanting activities, but instead some have voluntarily and proactively come forward to champion the implementation of some forest rehabilitation and replanting projects.

The implementation of Brunei HoB initiative can also be seen as an opportunity for Brunei Darussalam, especially the relevant government agencies, to promote a holistic approach in forest land use management. The establishment of HoB National Council of the 21 April 2008, with permanent members comprises of representative from Ministries which have relevancy over land use management, provides an important avenue for the relevant policy and decision makers to discuss and share ideas; exchange views and information; and streamlines projects and initiatives in order to reduce land use management conflicts and issues, hence ensuring the country to develop and grow in a sustainable manner, within the context of sustainable development.

An important point to note about this HoB initiative is that, while the initiative put its emphasis on forest conservation, however, this initiative does not intent to lock away development. Apart from the HoB initiative, there are also other policies, strategies, programmes and initiatives that are being implemented by the Department over the years, as explained in the following sub-headings.

Reduced-cut Policy

The reduced-cut policy, which was enforced since 1990, is part of the conservation effort that is in line with the 1989 National Forest Policy. With the reduced-cut policy, the annual logging rate was reduced by 50 per cent, that is, from 200,000 m^3 to 100,000 ^{3}. The shortage of supply to cater for domestic demand is offset by imported sawntimbers. This annual logging quota is still enforced until today.

Increase size of Forest Reserve

The 1989 National Forest Policy outlines the principles and strategies to, among other things, promote the conservation of biodiversity. One specific provision is the commitment of dedicating at least 55 per cent of the country's land area as forest reserves, that is, an increase of about 15 per cent from the 40 per cent that has already been gazetted.

Limit the issuance of Logging Permits/Licenses

The Forestry Department has also stopped issuing new logging permits or licenses since 1980's. This measure was taken as part of our conservation strategy in order to take into account on the limited areas of the production forest. To date, there are only 24 sawmills cum loggers still actively involve with the sawmilling and logging activities in the country.

Ban on the Export of Raw Logs

The law bans the export of timber. This is part of the country's forest conservation strategies in order to ensure sufficient supply of forest products for future domestic requirements.

Enhancement on Public Awareness

The Forestry Department has been and will continue to organize programmes that could promote and generate awareness among the public on the importance of conserving and protecting the forests as well as instilling the sense of love and appreciation for nature. Among the programmes that are being carried out annually by the Department is the celebration to commemorate World Forestry Day, which we normally follow-up with several forestry-related activities throughout the year. The activities include the mass tree planting for all walks of life, nature camp and nature excursion for the students as well as scientific research project for the secondary school students, that is, the prestigious Princess Rashidah Young Nature Scientist Award [PRYNSA].

Increase Forest Productivity

The Forestry Department will continue to intensify its silvicultural treatment activities in the logged-over forests in order to increase the productivity of the designated production forests. Well-planned forest harvesting system will continue to be practiced to ensure the silvicultural objectives of the natural production forest reserves are met. The Department will also continue to develop 30,000 ha of forest plantation by planting fast-growing species with crop rotation ranging from 15 – 40 years. These forest plantations are aimed at reducing dependence on the natural production forests. This was prompted by the projection that the natural production forests would be exhausted by the year 2015 if the logging rate were allowed to continue at 200,000 m^3 per year. The implementation of this forest resource development activity *i.e.* plantation establishment and silviculture treatment is one of the programmes under our 5 year series National Development Programme.

Establishment of Conservation Areas

The Forestry Department has also conducted programmes, which are in line with conservation activities, such as the establishment of *ex-situ* and *in-*

situ conservation areas; delineation of genetic resource areas; as well as germplasm collection, the later of which is conducted by the Agriculture Department. Excellent examples of *ex-situ* conservation collection in Brunei are those of selected tree species, palms, bamboo and rattans located at the Brunei Forestry Centre in Sungai Liang; and plant collections located at Forestry branch in Sungai Lumut. On the other hand, the Ulu Temburong National Park can be described as one of the largest in situ conservation area in Brunei. Situated mainly in the Batu Apoi Forest Reserve in Temburong District, the National Park is mostly a virgin jungle comprising various forest types classified according to altitudes and soil types. Apart from conservation, research, and education, the Park also caters to ecotourism.

Development of Brunei Tropical Biodiversity Centre [TBC]

One of the programme that has have been included under the current National Development Programme [NDP] is the development of the Brunei Tropical Biodiversity Centre. The main objective of the establishment is to ensure that conservation and utilization of biodiversity resources on a sustainable basis can be achieved. The main thrust of the centre focuses on providing opportunities for research, education, and eco-tourism. The master plan for the development of the Brunei Tropical Centre has been completed and now, under the current 2007-2012 NDP, actual implementation of the 1st phase of the project *i.e.* the construction of the TBC main building, is expected to be implemented by May 2010.

Strengthening International and Regional Cooperation

The Forestry Department has been very active in participating regional and international meetings, conferences, symposiums and workshops as well as strengthening cooperation with other research institutions, organizations etc. This effort is part of its commitment towards enhancing capacity building as well as facilitating the transfer of technology and sharing and exchanging information in forestry related fields, one of which includes the agenda on conservation of the forest resources.

CONSERVATION AND MANAGEMENT OF BIOMASS RESOURCES MANAGEMENT

Conservation and management of biomass resources and environment are embedded in an array of knowledge and local practices. Most development projects tend to ignore the ingenuity of rural people in responding to the local problems based on their local knowledge. Initiatives intended to grow trees through Social Forestry projects have not taken cognisance of the existing local practises. For example in the Himalayan region, common practices of obtaining biomass resources from forests are through pollarding, pruning or selective cutting and not by felling of tress. Similarly studies from Nepal have shown that, communities adjacent to forests mostly use dry wood as fuelwood and grasses for fodder that normally does not involve cutting of trees. For many local communities across South Asia, forests are not only of consumptive value, but also revered as places of great socio-cultural and religious significance. The other factors that regulate resource extraction from forests include (a) traditional rights (b) religious practices (c) and non-use of certain tree species in the region. For example, people have a strong belief that felling of certain tree species such as 'neem' (*Azadirachta indica*) and 'pipal' (*Ficus religiosa*) brings evil to the members of the household. Thus local systems of knowledge and management are sometimes rooted deeply in their religion and belief systems, which help in conserving the resources. But, state management was often based on flawed premises or a lack of understanding of the people-nature dependency. Gradually local communities have lost access and control of the resources, resulting in conflicts with the state and other stakeholders.

In many parts of south Asia, local people maintained strict rules to manage natural forests in their vicinity even until the 60s. The examples given below are illustrative of the diversity of informal or traditional institutional arrangements at the local level:

(i) A grazing fee was levied on herders in some pasturelands. In some regions, grazing was banned in pond catchments to prevent siltation of ponds.

(ii) Cutting of grass and fodder leaves from the certain tree species, 'Sal' (*Shorea robusta*) in the *Terai* region in the Himalayas, or 'Palas' (*Butea monosprema.*) in central and western India was regulated on forestlands.

(iii) Collection of dry wood did not require individual permits from the informal councils at the village level, whereas special permits were required for felling trees on the hill slopes and catchments. In case of some species that are less in abundance such as Shisham (*Dalbergia sisoo*) and Bamboo (*Dendrocalamus* spp.) there was a ban on felling. These species had market value for their timber.

The village community performed collective work to enhance and maintain the tree cover, by fencing, planting, trenching in forests, fire lines maintenance and desilting of ponds. Each household had an obligation to contribute a share to such efforts, failing this its rights to resources were curtailed. These duties are no longer followed as the local rights to resources are curtailed with state control.

The indirect non-consumptive values were represented in day-to-day life in the form of rules and conventions regulating use of resources. The informal institutions tend to reduce the environmental uncertainty faced by resource users especially in arid and semi-arid regions and also contribute to the conservation of natural resources. In some cases, the local institutions are better informed than the state departments about local ecological, economical and social conditions and about problems and constraints, which may be important from the management point of view. According to some studies, traditional management systems of CPRs have contributed to the protection of natural resources from being over used and have played a vital role in the management and conservation of natural resources.

TENURE RIGHTS AND BIOMASS RESOURCES

Prior to 1950s, local landlords controlled and had the authority to grant rights in forests in India, Sri Lanka and Nepal. In different areas, the tenure systems varied, for example the *Birta* system in mid hills of Nepal and the *Zamindari* system in northern part of India. The colonial regime in India and Sri Lanka significantly influenced the land tenure system.

Under colonial management systems which continued in some areas even during the postcolonial period, the local landlords had the responsibility to manage forests and granted rights to the local households. These include access rights, use rights and entitlement rights, which determine the access and end use of resources from the forests.

Timber extraction in general was regulated, but there was free access to non-timber forest products. These resources were treated by local communities in many parts of south Asia as a "free good" which were easily available. But these rights and duties ceased with the land reforms during 50s and subsequent development policies initiated by respective governments. In Nepal, forests were nationalized as a part of a wider move to break feudal structures. In India, it was the Land Revenue Act in the respective states, which was implemented to nationalize forests. Nationalization of forests did more harm to landless and rural poor whose access to common pool resources was restricted as part of these resources were declared protected.

In addition, communal sources of fuel wood, fodder and non-timber products have been reduced as a result of the forestry projects under the management and control of state functionaries. Due to nationalization of forest, the lack of local tenure and rights meant there was no incentive for local users, especially the poor and landless to exercise restraint. The rights to access and use of forest resources was further restricted with social forestry interventions enclosing forestlands.

For example, in the middle hill region of Nepal, the poorer sections of the people have less access to forest products for subsistence use and income today, than they had before the implementation of social forestry projects. This is also the case in India and Sri Lanka, where common lands, which were accessible to the landless prior to the social forestry intervention, are no longer freely available. In such a situation, the cultural overlay of rights creates paradoxically huge 'scarcity' differentials between households in villages, especially for the landless and poor. This tends to inhibit effective development and active participation by marginalized groups. Currently, the average poor household receives only one third to one fifth of the wood from forests than collected before the social forestry intervention. The rights to use resources from forestlands, which were formerly guaranteed by customary law, were banned after the land reforms and land acts came into existence in 1950s.

The choice of species or location of social forestry plantations was often made by people who had practically little local knowledge. The fact that there are local variations, even in small countries like Sri Lanka and Nepal, wherein women traditionally prefer certain tree species for fodder or fuel wood is important for any afforestation project to consider. The practice of use restrictions and the issue of land tenure, when viewed in the context of cultural and property rights based on gender inequalities and power relations within the households, presents a complex scenario. The tenure rights to land and resources in many South Asian regions are held by men, giving them legal rights over the resources, although women perform the frequent work of collecting biomass resources for household needs, cooking etc. Attached daily to the forest, women thus have a better knowledge on forest management and conservation, but are the most neglected group in social forestry projects.

There is no equity between men and women in land tenure specified officially in the land laws of the respective countries. Neither are women's rights in resource use are ensured in practice. Local traditions seems strongly influence on the women rights in land tenure, as it is the male child who inherits land. As a consequence, women's involvement in forestry projects is less, at the most they are being used as labourers in planting or watering the plants.

NUTRIENT CYCLING IN FORESTS

A nutrient cycle (or ecological recycling) is the movement and exchange of organic and inorganic matter back into the production of living matter. The process is regulated by food web pathways that decompose matter into mineral nutrients. Nutrient cycles occur within ecosystems. Ecosystems are interconnected systems where matter and energy flows and is exchanged as organisms feed, digest, and migrate about. Minerals and nutrients accumulate in varied densities and uneven configurations across the planet. Ecosystems recycle locally, converting mineral nutrients into the production of biomass, and on a larger scale they participate in a global system of inputs and outputs where matter is exchanged and transported through a larger system of biogeochemical cycles.

SOIL CHEMISTRY AND NUTRIENT VYCLING

By and large, the study of soil chemistry has concentrated on agricultural soils. The immediate practical questions to be answered have centred on the availability of plant nutrients, the need to develop chemical tests to define those soils suitable for particular crops and the prescription of the quantity of fertilizer that must be added to increase the productivity of the crop. With the exclusion of N (the availability of which is driven by biological processes in all soils), we might summarize the essential difference between the availability of plant nutrients in agricultural and forest soils:

- The availability of nutrients in agricultural soils is an immediate concern that mostly addresses inorganic equilibria over weeks or months;
- The availability of nutrients in forest soils is a long-term concern that should address biological processes and inorganic equilibria over years and centuries.

The growth of forests is a long-term process in which nutrients are cycled from plant to soil in litterfall and root turnover. Nutrients are withdrawn from tissues as they age and are translocated to actively growing tissues.

Timber harvesting takes away the aged tissues with relatively low nutrient concentrations, leaving the tissues with relatively high nutrient concentrations on site. In contrast, all of the nutrient supply for an annual crop is taken up from the soil within the few months the crop takes to reach maturity.

There is little cycling of nutrients from plant to soil. At maturity, the most nutrient-rich parts of the crop (seeds, leaves, storage organs) are harvested, leaving only the tissues with relatively low nutrient concentrations behind.

Thus, from the pioneering work of Ebermayer (1876) more than a century ago, much of the work on forest soils has been directed towards nutrient cycling, in particular:

- Quantifying the cycle of nutrients (uptake and return between plant and soil and retranslocation within the plant);
- Quantifying inputs and outputs, including nutrient removals in harvested timber.

While all of this is obvious, the science required is not obvious. Twenty years ago Stone (1979) concluded that despite a wealth of data on nutrient cycling processes for many forests, we do not have a sound experimental basis on which to assess the sustained productivity of forest ecosystems. Ten years later, Landsberg *et al.* (1991) restated that conclusion: 'Conventional chemical analyses to determine the nutritional status of the soil yield information that may have little relevance in calculating the capacity of the soil to provide nutrients to trees, and the complexity of uptake processes through complex, dynamic and poorly defined root systems is tremendous.'

Rather than attempting to review ongoing research on the experimental and analytical basis for assessing sustainability of nutrient supply, we give only a factual account of what is known of the chemistry and cycling of the major nutrients. We do this in general terms, using data for a temperate forest of above-average productivity and we include a summary of the form, function, concentration and cycling of the major nutrients N, P, Ca, Mg and K.

Biomass

The above-ground mass of the forest (trees, under-storey and shrub layer) may reach 500t ha^{-1}. Litterfall *(A)* is 8tha^{-1} $year^{-1}$ but may reach 10tha^{-1} $year^{-1}$. If the litter layers (*Q*) weigh 30tha^{-1}, the decomposition rate of litter $k = A/(A + Q) = 0.21$ $year^{-1}$, and the half-life $t_{0\ 5}$ = 0.693/ft = 3.3 years. This and subsequent analyses are based only on the above-ground stand. Data for root mass and root turnover are both more limited and variable, and have been intensively summarized. We have previously estimated a rate of root turnover for our temperate forest of above-average productivity of 2-3tha^{-1} $year^{-1}$.

Nitrogen

The concentration of N in leaves is greater than the concentration of any other nutrient, but much of this N is withdrawn from leaves before they die and fall as litter. Internal redistribution of N accounts for almost 30 per cent of the total annual above-ground cycle. The half-life of N during litter decomposition is 3.5 years, a little longer than that for dry weight, so that N is immobilized during decomposition of the litter. The input of N in rainfall of 5 kg ha^{-1} $year^{-1}$ is more or less balanced by the output of N in streamwater. More than 95 per cent or so of N in the soil is in organic form; organic N is mineralized to NH_4^+-N and then under certain conditions to NO_3--N, a form freely available for uptake by plants. NH_4^+-N is held on negatively charged exchange sites of soils, but NO_3--N is mobile within the soil and is the form in which N moves with drainage waters into streams.

In soils of natural forests of the world, the generally high C/N ratio of litter and soil organic matter precludes mineralization proceeding to NO_3--N, so that the dominant form of N is NH_4^+-N. The cycling of N is therefore conservative, and N may be immobilized in litter of older forests (particularly in cooler-temperate areas) to the extent that the trees become N-deficient. Disturbance (treefall and the creation of gaps, fire, harvesting for timber) provides the conditions that result in a decrease in C/N ratio and the production of NO_3--N, in which form N may be leached from the soil into streamwater. Rapid regeneration of the forest, including rapid uptake of NO_3--N, following disturbance is therefore the key to the conservation of N.

Trace Elements (micronutrients)

The trace elements Fe, Mn, Cu, Zn, Mo and B have all played spectacular parts in agriculture and in tree plantations where plants have been introduced into new land. Applications of trace amounts (*e.g.* as little as 0.5kgha^{-1} of Mo) have corrected deficiencies that would otherwise have made agriculture or plantation forestry unproductive. We know of no such experience in natural forests, although there is little doubt that the distribution of many species may be determined by their ability to increase or restrict the supply of one or more of the trace elements. For example, concentrations of Mn in leaves and bark differ consistently and significantly between groups of the genus *Eucalyptus* but whether these differences are of significance in determining the distribution of species is unknown.

Phosphorus

Phosphorus occurs in highly mobile forms within plants and, like N, is withdrawn from plant parts during senescence; internal redistribution of P accounts for 50 per cent of the total annual above-ground cycle. The half-life of P during decomposition is 4.1 years, *i.e.* P is immobilized throughout decomposition to a greater degree than N. A negligible amount of P comes in with the rain. P as H_2PO_4- is adsorbed by clay colloids and is then highly immobile in the soils so that losses of P to streamwater are negligible. However, in forest soils more than 50 per cent of the total P in surface soils may be in organic form. Mineralization of organic P is therefore fundamental to ecosystem function, and P availability is 'determined by competition between biological and geochemical sinks for phosphate anions'. Whereas methods for estimating rates of N mineralization have been developed and are in general application, there are no methods for routinely estimating the rate of P mineralization. The difficulty is that whenever H_2PO_4- comes into solution, there is a surface to adsorb it. This competitiveness between biological and chemical sinks for P remains a priority area of research in quantifying the long-term sus-tainability of P supply for forests.

Calcium, Magnesium and Potassium

Ca, Mg and K, and the non-essential element Na, are generally grouped as the major exchangeable cations in soil. An exchangeable cation is held adsorbed on a negatively charged surface of a colloid (both inorganic and organic). In this exchangeable form, the cations are not soluble but are brought quickly into solution and made available for uptake by exchange of protons from plant roots.

Ca is highly immobile within the plant, where it is incorporated within cell walls. Thus 80 per cent of the total annual above-ground cycle of Ca is in litterfall. In contrast, K is highly mobile and K^+ easily leached from tree crowns by rain; leaching accounts for about 30 per cent of the total annual above-ground cycle of K.

Roots and the Rhizosphere

Roots supply plants with water and nutrients and provide anchorage in soil. Most studies of roots in forests have been directed to the surface horizons, where fine roots (< 2 mm in diameter) are concentrated and where water and the bulk of nutrients are sourced. There are few studies of roots below 1 m, with only nine reported in the review of Jackson *et al.* (1996). In general our knowledge of root characteristics, including biomass, distribution and fine-root area, is poorly developed compared with the above-ground components of forests. Much of the recent interest in root distribution and turnover in forests has arisen primarily from concern for increased C flux from soils due to climate change.

Distribution, Mass and Turnover of Roots

A global analysis of root distribution in terrestrial biomes showed that boreal forests have the shallowest rooting profiles (80-90 per cent of roots in the upper 30 cm of forest soils) and that temperate coniferous forests have the deepest rooting profiles (52 per cent of roots in the upper 30 cm). When considered together, all temperate and tropical trees have 26 per cent of roots in the top 10 cm and 60 per cent in the top 30 cm. Root biomass of forests has been estimated at 2-5 kg m^{-2}, with the highest values in tropical evergreen forests. However, discrepancies between model estimates and measured characteristics of fine-root length and surface area point to the need for more data on root biomass and distribution, and for more careful measurement of fine root biomass in particular.

These analyses of root distributions and characteristics (biomass, surface area and nutrient content) among forests will enable improved modelling of water and nutrient uptake on a global scale. They also provide the basis for improving global models of the impact of climate change on vegetation distribution and C sequestration in soil.

When linked with models that predict the change in vegetation distribution and type, they will allow better prediction of the consequences of global environmental change. The need for this type of analysis seems all the more urgent when it is considered that roots can account for more than 50 per cent of net primary productivity (NPP) in forests and that the global pool of C in fine roots is 5 per cent of the size of the atmospheric C pool. Because about half the global pool of C in fine roots is in forests, further study of factors influencing fine-root turnover is critical to the understanding of CO_2 flux between terrestrial and atmospheric C pools.

While roots are concentrated in the upper metre of soil in most forests, deep roots become important in forests subject to periodic drought because they provide access to water held deep in the soil. Carbon from these deep roots may contribute over decades to net C release from soil following conversion of forest to pasture, as shown by Nepstad *et al.* (1994) in an Amazonian rain forest. These sources of deep soil C have not been generally considered in estimates of C flux from forest soils, especially those of tropical regions where forests continue to be cleared for agriculture.

The turnover of fine roots has been investigated in relation to forest disturbance, including clearcutting, in the Northern Hardwood ecosystem. Biomass of fine roots (<2mm) in mature hardwood forest was 471 gm^{-2}, with average lifespan ranging from 8 to 10 months. Biomass of fine roots recovered to 71 per cent of that in the mature forest within 3 years after clearcutting.

Rhizosphere

The rhizosphere is the area of soil immediately surrounding the root where biotic and abiotic processes interact to create an environment distinct from bulk soil further from the root. Compared with bulk soil, rhizosphere soil is usually more intensively weathered, has a lower pH and has greater concentrations of cations and P. Exudates released from plant roots and microorganisms benefit plant growth by increasing nutrient acquisition and metal detoxification, by alleviation of anaerobic stress in roots and by their action in mineral weathering.

There is no doubt that some forest species can modify the rhizosphere to access previously unavailable reserves of soil P by releasing organic acids that solubilize inorganic P. For example, Grierson (1992) demonstrated that in P-deficient soils, *Banksia integrifolia* induced the development of short-branched, tertiary, lateral roots (proteoid or cluster roots) that released citric and malic acids (the main organic acids induced under P deficiency). Organic acids probably increase the availability of P in soils through a combination of decreased adsorption of P, increased solubiliza-tion of P compounds and chelation of metals such as Fe. Other rhizosphere processes that impact on the availability of soil P reserves include increase in root hair length and density, release of C that enhances mycorrhizal exploitation of bulk soil, and the release of phosphatases that solubilize organically bound P.

The role and function of mycorrhizas in forest soils has been reviewed by Brundrett (1991) and Vogt *et al.* (1991) and there is no doubt that mycorrhizal associations benefit trees by enhancing nutrient acquisition, especially in relation to P. The evidence for N and P transfer to forest trees via mycorrhizal associations has been demonstrated in tracer studies run over relatively short time-courses. However, it is only recently that new techniques have allowed longer-term studies; these have confirmed the benefits of mycorrhizas to nutrient acquisition and have also demonstrated increased plant growth as a result. Further work of this kind is required to establish the relative contribution of mycorrhizal uptake mechanisms to nutrient acquisition in mature forests over the long term.

The activity of microorganisms in the rhizos-phere is driven by the release of low-molecular-weight organic compounds by the root and there is some evidence to suggest that the growth of certain microorganisms is favoured while that of others is suppressed. For example, Gonzalez *et al.*(1995) reported that while microbial numbers were greater in the rhizosphere of *Alnus,* roots differentially favoured colonization by proteolytic and ammonifying organisms and inhibited nitrifying organisms. Norton and Firestone (1996) reported a greater than 50 per cent increase in N turnover (mineralization and immobilization) in the rhizosphere relative to bulk soils in *Pinus ponderosa* microcosms. Further evidence of the complex interactions in the rhizosphere come from reports of ectomycorrhizal fungi stimulating bacterial growth associated with the release of citric acid by the fungus and, conversely, of inhi bition of bacterial growth.

Recent work has aimed at characterizing the chemical and biotic nature of the rhizospere, especially as it is influenced by soil liming and nutrient addition. Most work has been focused in the vicinity of industrialized areas of the Northern Hemisphere, where there are concerns for the unimpeded functioning of beneficial root-rhizosphere reactions under both controlled (fertilizer inputs) and uncontrolled (atmospheric deposition) inputs of acidity, N and S. Reported effects of ammonium sulphate application on rhizosphere chemistry vary and many of the changes, such as increased acidity and Al, are found in bulk soils as well. Other effects include depletion of P and K from the rhizosphere of Norway spruce in southern Sweden induced by ammonium deposition and the subsequent stimulation of P and K uptake.

The benefits of rhizosphere organisms, including mycorrhizal fungi, in maintaining the growth of tree seedlings in nurseries has long been known. This concept has been extended recently to the development of soil inoculation techniques to improve the growth of seedlings in the reforestation of degraded or clearcut sites, where inoculation with forest soil has increased the number of ectomycorrhizas. Differential effects on survival and growth of seedlings, where planting holes were inoculated with soil transferred from forestry or plantation sites, have been reported. Increases in seedling growth have been associated with increased nutrient mineralization brought about by soil animals stimulating microbial turnover in transferred soil, while increases in survival probably resulted from the introduction of beneficial rhizosphere organisms.

Another dimension in the use of mycorrhizas to benefit tree growth has been the recognition by Garbaye (1994) of bacteria that selectively promote mycorrhizal development (the so-called 'mycorrhization helper' bacteria, MHB). In anticipation of the applications discussed by Chanway (1997) the process of selecting and using MHB cultures for controlled 'mycorrhization' has been patented by Garbaye and Duponnis (1991).

8

Forest Protection and Management

INTRODUCTION

Agencies that seriously damage and may destroy forests include fire, insects, diseases, grazing animals and adverse conditions of exposure and climate. The relative importance of these varies with local conditions.

Fire is perhaps the most spectacular. It frequently destroys not only the forest growth, but also the surface organic layer beneath, thus reducing soil fertility. Fire is most serious in the coniferous forests of temperate regions and in the dry forests of the tropics and subtropics and is of course most difficult to control in heavily forested regions with sparse populations and poor accessibility.

Protection against fire is usually one of the first steps taken in the development of a forestry programme. Appropriate legislation is passed and organized staff and equipment and transportation facilities are provided to detect and suppress fires as they occur. Where the country is heavily and continuously wooded, strips may be cut through the forest to break the continuity of tree growth and delay or prevent the spread of fires. Clear-cut strips of this nature are known as "fireguards." Fire itself is sometimes used as a preventive measure where the annual fall of litter and the dying back of ground vegetation produce large accumulations of inflammable debris. The "early burning" of this material when moisture conditions are such as to prevent too rapid spread of fire will reduce the danger during a dry period later in the season.

Recent important advances in the field of fire protection include the use of aircraft, mainly for detection and of meteorological data to appraise more precisely the existing degree of fire hazard, so that dangerous conditions can be quickly recognized and adequate preventive measures taken.

In some European countries forest fires are rare and such as do occur cause negligible damage. In Norway, for example, forest properties may be insured against fire at a low premium rate. There are a number of reasons for this: *(a)* in general the forests are easily accessible and under intensive management, with annual fellings well distributed in a large number of small scattered holdings; *(b)* the forest consciousness of the people is highly developed; *(c)* the climatic conditions tend to be favorable, with well-distributed summer rains and an absence of prolonged hot, dry periods.

Despite the damage it causes, fire cannot be looked upon solely as an enemy of the forest. Under certain conditions controlled burning may prove a useful tool in securing the regeneration of valuable forest trees; it may prepare a suitable seedbed, eliminate competing vegetation, or assist in the distribution of seed. Certain conifers, for example, have serotinous cones and considerable heat is required to open them and release the seed: these species may reproduce abundantly after fire and the resulting stand is often referred to as a "fire type." Extensive even-aged coniferous forests of the northern temperate region owe their origin to fire, as do many other commercially important forests in various parts of the world.

Insects that damage and destroy trees and the fungi and other organisms that cause tree diseases, are present in all forests. In endemic proportions and in balance with their environment, they are a necessary and sometimes innocuous feature of the forest community. Even under these normal conditions, however, some insects and diseases may seriously damage commercially valuable trees. While the damage caused by fungi is usually slow in developing and of minor account in young trees, it is cumulative and may be so extensive in overmature trees as to render them valueless.

Destructive organisms may also reach *epidemic* proportions and destroy one or more tree species over wide areas. This is particularly likely to happen in overmature forests or when an organism is moved from its natural environment to new surroundings.

A classical example is the chestnut blight *(Endothia parasitica),* which was brought to North America from Asia about 1900 and killed practically all of the sweet chestnut *(Castanea dentata)* in eastern America within a few years. Instances may also be cited of widespread destruction by forest insects, such as the spruce budworm *(Choristoneura fumiferana)* in the spruce and balsam-fir forests of northern and eastern Canada.

Methods used to control insects and disease include the removal of trees most susceptible to attack, the establishment of new stands containing a mixture of species rather than pure stands and discrimination against species susceptible to an insect or disease known to be present in the locality. For example, white pine *(Pinus strobus),* which is often badly attacked by the pine weevil (Pissodes strobi) and the blister rust *(Cronartium ribicola),* should not be encouraged where these pests are commonly found, unless some effective means of control is available.

Other methods of controlling forest pests include the introduction of parasites and predators, the use of chemical sprays and the development through selection and crossbreeding of strains or hybrids that are relatively immune to attack by specific pests. Domestic animals grazing in forest areas, particularly cattle and goats, often cause much damage by killing seedlings and compacting the soil. In the eastern Mediterranean region, where goats have been allowed to roam over the countryside, much of the forest vegetation has been completely destroyed. In many cases the only control possible is to exclude the animals; elsewhere forest grazing may be permitted under carefully controlled conditions.

Forests are normally protected against the adverse effects of weather and exposure by some adjustment in the methods of cutting and by encouraging the development of species best suited to local conditions.

REPRODUCTION

A tree crop may be established by artificial or natural means, or by a combination of the two. In artificial reproduction, seed is sometimes sown, but more commonly young trees that have been grown in nurseries are planted on the site. In either case, the seed used is collected in the forest, preferably from trees of good form and vigour. When the crop is established by natural means, the young trees may originate vegetatively or from seed disseminated either before felling takes place or by trees left standing for the purpose at the time of logging. A new crop of young trees that becomes established in the forest before cutting takes place is known as "advance growth."

There are two principal methods of cutting the mature forest, each of which has advantages with respect to the regeneration and establishment of individual tree species:

Clear-Cutting

By this method, which results in the development of more or less even-aged stands, the forest is cut in one operation or in a succession of operations, each of which removes either a portion of the trees scattered over the whole area or all of the trees from a part of the area; these operations are continued until the original stand is completely cleared. Regeneration may be established by natural or artificial means, the former being provided for either by advance growth, or by seed lying in the surface soil at the time of cutting or made available from the crowns of felled trees or from standing trees surrounding the cutover area.

For species that require overhead protection in their youth a "shelterwood cutting" is made: the stand is opened up gradually over a period of years by making three or four successive cuts periodically over the whole area until the original stand is removed. For other species, the seed source may be the critical factor and the stand is clear-cut except for a few mature trees, evenly distributed over the area, which are preserved to provide the seed for the next crop. This is known as "seed-tree cutting." Quite commonly, portions of a stand are clear-cut periodically in the form of narrow strips, wedge-shaped blocks, or patches, until the stand has been completely felled. The Douglas-fir forests on the Pacific coast of North America, for example, are clear-cut in patches or blocks of a size that will permit natural regeneration by seeding-in from the surrounding uncut stand.

This type of cutting is particularly applicable to species like the Douglas fir that are intolerant of shade. The intensity of cutting and the size, form and arrangement of clearcut areas vary according to local conditions and requirements, particularly stand density, topography, logging methods and the ecological characteristics of the desired species.

Selection Cutting

This method distributes cutting and regeneration continuously over the whole area, producing a single uneven-aged stand or series of such stands where trees of all ages from seedlings to mature timber occupy the area together. Scattered single trees or small groups of trees may be removed periodically throughout the whole forest, or the forest may be divided into a number of blocks, one of which is cut over in this way annually or every few years.

The aim is to improve and regenerate the forest, as well as to harvest a crop: while some of the better formed mature trees may be kept in the stand for a number of years to provide seed, all dying and defective trees that are at all merchantable are removed as quickly as possible. Non-merchantable trees may be girdled. As regeneration and early development of growing stock must take place in partial shade, this method is particularly suited to tolerant species.

AFFORESTATION

Forestry not only seeks to ensure the reproduction of existing forests, but is also concerned with the problem of establishing new forests. Afforestation may be undertaken *(a)* to reclaim forest land cleared for other purposes but best suited for the growing of forest crops, *(b)* to establish forests as an economic crop on land not previously forested, *(c)* to stabilize shifting sand and eroding soils, *(d)* to provide protection for agricultural and other crops against wind and snow damage and *(e)* to improve seepage in drainage areas and so stabilize stream flow.

Great Britain, Australia, New Zealand and South Africa have all undertaken large-scale afforestation with the object of supplementing their natural resources. The trees used have been predominantly softwoods, particularly species of pine. One of the best-known examples of land reclamation for forestry purposes is in the Landes of Gascony, where, since the latter part of the eighteenth century, a programme has been developed to fix shifting sand dunes, to drain swamps and to afforest large areas with maritime pine the principal source of resin and turpentine in Europe. The use of trees in shelter belts and windbreaks to protect agricultural land is common practice in many countries and has been practiced on a large scale in the United States and Russia. The establishment of forests to conserve water supplies has often been an important objective of forestry ever since its early beginning in Europe. In many cases, however, such forests may be used for wood production as well as for protection. In India and certain other tropical countries forests created primarily to provide small timber and fuel for the local population often help to solve the problems of flood control, erosion and desiccation.

The success of afforestation programmes depends upon many factors. Foresters must determine the suitability of species to soil and climatic conditions; they must guard against the many hazards that may arise in introducing a species or race to a new environment; they must decide whether the species grows best in a pure stand or in mixture with other species and whether it will thrive as a pioneer under open-grown conditions or is better suited to a later stage in plant succession. The relationship of the species to its environment must be studied carefully if afforestation is to be successful.

"TENDING" OPERATIONS FOR YIELD AND QUALITY

Once the forest is established, the aim of the silviculturist will be to develop and maintain the optimum growth and quality of the stand in keeping with the objectives of management. This is accomplished mainly by cutting operations of various kinds and intensities, especially by means of thinnings carried out periodically throughout the life of the stand. These are commonly referred to as "tending" operations, as distinguished from those aimed primarily at regeneration.

Although forest trees are members of a community and, as such, contribute something towards each other's well-being — *e.g.*, mutual protection against the elements and enrichment of their common habitat, the forest soil — there is, as noted above, a continuous and sharp competition among them for light and space and in some cases for soil moisture and mineral nutrients. The stand density (number of stems per acre) greatly affects the intensity of this competition and also influences the rate of growth per tree and the ultimate value of the forest to man.

Crowded conditions and keen competition are usually beneficial in the very early life of the stand: the inherently vigorous individuals become dominant, the lower branches of the trees remain small and are killed back at an early age and the tree tends to develop a straight bole.Later, however, too much crowding is a disadvantage. With reduced incidence of sunlight, radial development of the crowns is restricted and growth is slowed down.

In fact, stagnation may take place, especially in species showing little differentiation in height growth. It is the aim of the silviculturist to make the most of early competition while this is advantageous and to remove it when it hinders stand development — in other words, to distribute the total productive capacity of the site among the optimum number of trees to obtain the desired results. This is usually accomplished by thinning operations.*Thinnings.* Cuttings made in a fully stocked to overstocked immature stand in order to increase the rate of growth of the remaining trees are known as "thinnings." In general, these are begun in the late sapling stage and are continued periodically until the stand matures: they are usually of only moderate intensity until rapid height growth is completed, the later thinnings being more severe to accelerate diameter growth.

The total annual growth of an area is not materially changed by thinning, but it is confined to fewer trees whose individual increments are thereby increased. As a result, the trees reach useful sizes at earlier ages, which is particularly important when logs of large size are desired for sawing into lumber. Too heavy a thinning, however, may leave insufficient trees for full use of the productive capacity of the site.The economic feasibility of thinning operations is largely determined by the availability of markets for small-sized material, although in some cases the improved growth of the residual trees may justify the operation.*Other Cultural Treatments.*

To promote more desirable species and better formed trees and generally to improve the quality of the stand, other cultural treatments may be used, among them cleanings, liberation cuttings and improvement cuttings.

- *Cleaning.* An operation in a young stand not past the sapling stage:
  - To free small trees from weeds, vines, or sod-forming grasses;
  - To provide better growing conditions by liberating crop trees from other individuals of similar age but of less desirable species or form that are overtopping or likely to overtop them.
- *Liberation Cutting.* The release of young trees not past the sapling stage from competition with older trees that are overtopping them.
- *Improvement Cutting.* A cutting made in a stand past the sapling stage to improve its composition and character by removing trees that are less desirable with respect to species, shape, or main crown canopy.

Pruning. Knots in lumber are the result of branch growth; the fewer and smaller the branches, the fewer and smaller the knots and the higher the grade of lumber produced. In a young stand of proper density the lower branches are often small in diameter and die at a fairly early age.

The forester may assist in the production of knot-free wood by removing the dead and dying branches from the bole of the tree with axe or knife. Care must be taken, however, not to injure the tree nor to reduce unduly the living portion of the crown. A comparison of the cost of pruning with the relative value of knot-free wood will determine the feasibility of this treatment.

SILVICULTURAL SYSTEMS

As a result of past study and experience, particularly in northwestern

Europe, there has been developed a series of silvicultural systems, each incorporating the techniques that are required in the silvicultural management of a specific type of forest to maintain its distinctive form and maximum productivity.

While these systems are based primarily on the method of regeneration (whether by coppice or seeding and with reference to the form of cutting), they are also concerned with problems affecting the protection and tending of the forest and the harvesting of the crop and its economic utilization. Where such systems have been evolved, they will form an essential element in the plan of forest management.

PARTICIPATORY FOREST MANAGEMENT

With the Pace of Population booming, increased energy consumption, over exploitation of the natural resources and rapid depletion of the forest reserves accelerated natural disaster like flood, drought and cyclones in Bangladesh.

Once Bangladesh was famous for its evergreen/ semi evergreen tropical and world famous mangrove forest. But over the years due to over exploitation of forests and its non-participatory management, more than 50 per cent of the forest resources has been depleted. Realising the grim effect of destruction of forests and to repair the lapidated environmental condition, both the government and non-government organization have taken up afforestation programme. The NGOs have added a new dimension in the forest management, which has ensured participation of the community people and protection of the vegetation. Although, the government has also adopted participatory forest management but due to bureaucratic attitude easy access of the poor habitants are restricted in many cases. To overcome these situations, the existing government forestry policy, which was formulated in 1994, needs radical modification. There should be room to accommodate the NGOs, grass root organisations and general people in policy formulation, execution and evaluation of the programme.

Bangladesh lies in the North-Eastern part of South Asia between 20°34' and 26°38' North in latitude and between 88°l' and 92°41' East in longitude. The total area or the country is 144,000 sq. km with a population of about 120 million, density of population is 800 person per square kilometer. The most densely populated country in the world, Bangladesh is mainly a floodplain delta, which is formed at the confluence of the Ganges, the Brahmaputra and the Meglina rivers. Natural forest represents only 6 per cent of the total land area of the country and is managed and controlled by the government. Village forest which contains annual and perennial trees and still provides a major source of food and income for majority of people, is managed by private individuals.

A rapidly increasing population is placing growing demand on natural resources, especially forest sector is under pressure to become more productive and efficient to keep pace with increasing demand. At present forest in

Bangladesh is in unfavourable situation in terms of meeting increasing demand and also not adequate for maintaining ecological balance. This is primarily due to heavy population pressure and limited resource base, secondly, lack of integrated planning for development of multiple resource base with active participation of people resulting in high degree of environmental degradation, as illustrated mostly by deforestation and destruction of Natural resources.

In the phase of rapid depletion of forest aggravated by increasing demand for forest resources, and considering the prevailing socio-economic condition of the country, Government has put emphasis on participatory approach in development of forest resources of the country.

FOREST SITUATION IN BANGLADESH

Bangladesh has lost over 50 per cent of its forest resource over the period of about 25 years. Actual forest coverage is only 6 per cent of the total area and the situation is worsening despite of an attempt to preserve it. At approximately 0.02 ha per person of forest, Bangladesh currently has one of the lowest per capita forest ratio in the world. In Bangladesh, government-owned forest area covers 2.19 million ha, with the remaining 0.27 million ha being privately controlled homestead forests. Of the government owned forest land, 1.49 million ha are national forests under the control of the Department of Forest, with the rest being under control of local governments. Of the state owned forests, over 90 per cent is concentrated in 12 districts in the Eastern and South-Western region of the country. However, due to over exploitation these forests have become seriously degraded. The natural forests of the country are classified into three categories: 1) Tropical evergreen/ semi-evergreen forest in the eastern districts of Sylhet, Chittagong, Chittagong Hill Tracts, and Cox's Bazaar: 2) Moist/dry deciduous forest also known as Sal forests in the central and the northwest region and 3) Tidal mangrove forest along the coast, known as the sundarban, the largest mangrove ecosystem in the world. These forests are official reserves and placed under the jurisdiction of the Forest Department. Unfortunately, recent inventories indicate a continuing depletion of all major forests.

Forest management in Bangladesh

In Bangladesh management of government forest is the responsibility of the Forest Department under the Ministry of Environment and Forest. In this process the department is managing, protecting, developing the forest resources, forest land and also collecting the revenues. People have never been consulted nor involved in forestry activities. From the management point of view, forest of Bangladesh are being divided into three categories such as:

- State owned forest under the administrative control of Forest Department.

- State owned forest under the administrative control of Ministry of Land through District administration.
- Private village forest managed by private individuals. Forest under Forest Department control and management again divided into three major types viz; (a) Hill Forests; (b) Plain land Sal Forests, (c) Mangrove Forests.

Hill Forests: The tropical evergreen/semi evergreen forest cover as approximately 1.32 million ha of which 0.67 million ha is controlled by the forest department and rest is under the control of hill district council. Clear felling followed by replanting with suitable species (both long and short rotation) is the method of management in hill forest. Because of increased demand for timber and fuel wood and prevailing socio-economic condition of the country this forest has greatly affected and rate of denudation is considerably high. The forest department is mainly confined in raising of single species plantation. Inventory shows that most of these plantations would not give the desirable output. This programme suffers from technical, social and administrative soundness. Another problem is most of the high forest are subjected to shifting cultivation by the hill tribes. The tribes are entitled to shifting cultivation in forest land under administrative control of district administration which has resulted in the total destruction of these tropical evergreen forest. The growing stock has depleted from 23.8 million m^3 in 1964 to less than 20.7 million m^3 in 1998.

Mangrove Forests: Known as Sundarbans, the largest mangrove ecosystem in the world. Sundarban forests are being managed by selection felling method followed by natural regeneration. Beside Sundarbans, plantations are being raised with mangrove species in the newly accreted char land all along the Coast of the Bay of Bengal. Sundarban forest is an official reserve forest, unfortunately recent inventory shows a continuous depletion due to over-cutting, illegal felling. It is estimated that in less then 25 years, the volume of commercial species Sundari, Gewa, has declined by 40 to 50 per cent respectively.

Plain land Sal Forests: Silvicultural system applied for Sal forest was coppice with standard system.

In this system matured trees were felled and the areas were protected for coppice regeneration. The typical nature of Sal forest is that this forest is scattered. In the forest areas there are agricultural lands owned by the adjacent people.

Frequently these land owners are extending their lands and encroaching to forest and in the process they are destroying the forest and subsequently converting the area to agricultural land. In this process forest lands are being marginalised day by day. FAO estimated that only 36 per cent of the Sal forest cover remained in 1985; more recent estimates that only 10 per cent of the forest cover remains due to over exploitation and illicit felling through there is an official base on logging since 1972. Most of the Sal forest are now substantially degraded and poorly stocked. The situation calls by for involvement of community people in the forest management.

Inventories how that there has been overall depletion in forest resources in all major state owned forest. The growing stock in Sundarban has been depleted from 20.3 million m^3 in 1960 to 10.9 million m^3 in 1998. In the Hill forest of hill districts, the growing stock has depleted from 23.8 million m^3 in 1964 to less then 20.7 million m^3 in 1998. Over-cutting by timber merchants, increased consumption linked to population growth, shifting cultivation, encroachment, illegal felling and land clearing for agriculture, lack of participatory management have been the principal causes of deforestation and shrinking of forest land in the country.

Since 1960 two major approaches regarding the role of forestry in development have been reflected in the forestry sector of Bangladesh. In the 1960's, Bangladesh as a part of Pakistan and then as an independent nation has followed 'An Industrialisation Approach' consonant with the international conventional wisdom at that time. As a result, Department of Forest raised large-scale Industrial plantation which were seen as conversion of low-yielding natural forest into artificial plantation of species (mostly teak) of great economic importance. This conversion of semi-evergreen and evergreen forest into deciduous teak plantation was largely concentrated in hill forest areas. During the plantation raising local people were not consulted and often they did not drive any benefits from these plantations. The lack of support by the local people/ communities in combination with lack of silvicultural knowledge and lack of proper maintenance contributed to raise low quality plantations and these plantations were also lost due to illegal felling. In the name of plantation the genetic resource of the ever-green/ semi-evergreen forest was lost. Forest Department was considered as revenue earning department. The main activities of Forest Department were concentrated in extraction of trees from the forest and replanting of those felled areas where applicable, Forest Department has not considered the people and their participation in managing forest of the country.

In the 1980s following a change in thinking about the role of forestry in development, and peoples participation in forestry activity was encouraged. People participation with the forestry sector realised the need of people oriented forestry programme to replenish the degraded forest resources of the country. Accordingly, in 1994 Government formulated a forest policy replacing earlier one enunciated in 1979 with a due emphasis to the need for people's participation in forest management.

ECONOMIC PRODUCTION IN COMMERCIAL FOREST MANAGEMENT

In order to get the most beneficial economic production out of our local forests, the practices and techniques of forest management need to be applied. As of 1924, Wicomico County was roughly 46 per cent wooded with the other 54 per cent covered by agricultural, commercial, and residential land. Most of the 46 per cent wooded land was and has been privately owned. In Wicomico County, the majority of wooded areas are privately owned and are mostly idle lands on a landowner's property. As a landowner, especially a farmer, the more economic profits you can get for your land the better. During the late 19th century and early to mid 20th century farmers and landowners were managing their idle lots into pine plantations. The February 17th, 1948 edition of the *Salisbury Times* even had an article entitled "Farmers Discover Timber is a Crop, Shore Benefits Too" and in this article it explains how modern day forestry practices are encouraging the re-growth of loblolly pine plantations in active timber lots. The dominant forestry management practices during the 20th Century were commercially based, meaning most of our forests were used to support timber and a very small portion of Wicomico's forests were being conserved to support diversity. Through different forestry management practices to promote commercial forests there has been a major loss of ecological diversity in Wicomico County.

PRESCRIBED FIRES IN FORESTRY MANAGEMENT

One forestry management practice was borrowed from the local Native Americans and used to protect the commercial loblolly dominated forests. During the 1920's through the 1930's the United States and especially Maryland went through a dry era. This was also around the Great Depression and the famous Dust Bowl in the Mid-West.

The pine industry in Wicomico County was not only an important industry it was an economic resource that affected most of the citizens in the county. Devastating forest fires became a threat to dry pine plantations and the fires would consume thousands of dollars in timber. The reason why these fires were so devastating is because dead vegetative matter including layers of dead pine needles and cones as well as other dead vegetation was allowed to accumulate on the forest floor. This built up matter that would naturally dissipate due to natural forest fires was not allowed to because fire was viewed as a costly event and forests were highly protected.

In the 1950's foresters figured out that fire could be the best tool to prevent more fires and to help promote the loblolly. Like thousands of years of fire forest management practices used by Native Americans, prescribed burns were used in Wicomico to prevent large costly fires from happening. Prescribed burns are man-made controlled fires applied to pine plantations used to burn up extra forest floor material and to eliminate unwanted plants, trees and future fires. Another positive of a prescribed burn is that the burnt material gives the soil back vital nutrients that help promote the growth of more trees.

Environmentally the prescribed fires of the 20th Century took its toll on diversity. The Natives used fire to control the forests to be diverse to support hunting. The fires during the 20th Century were being used to not only burn excess forest floor matter but to also eliminate competing vegetation of the loblolly pine. These fires eliminated young hardwood saplings from maturing as well as other shrubs that many species depend on. Also the smoke from prescribed fires was not too pleasing to the local communities. With the right winds the smoke would drift to people's homes, along highways and into towns making many people unhappy. One of the last things a person would like to smell on a nice spring day is smoke.

Clear-Cut/Regeneration Harvests

Clear-cutting, also know as the regeneration cut, has become the most economically beneficial harvesting management technique for the loblolly pine. Before forestry management existed, clear-cutting was used to open up the land for agricultural uses and to make settlements for the early settlers. Many years after these fields were abandoned the first trees to reemerge were the loblolly pines because they grow well in direct sunlight.

Loblolly forests then began to make up most of the county while the diverse mixed hardwood/softwoods stayed in the low-lying, wet, untouched areas. While this nation was growing, it needed a timber supply to help construct cities, boats, and crates and to fulfill other construction requirements. Wicomico County and the timber industry, were extremely important because of its fast renewable timber resource and accessibility to major cities on the eastern sea board.

Clear-cutting was one of the first management practices used. It was used because it allowed the industry to get as much timber as possible in a short period of time. Clear-cutting is selecting a wooded lot and cutting down all the trees and clearing away the brush. This method of forest management works well with the loblolly because its' seed cones can naturally regenerate new trees and the open sun is great for their growth. As Wicomico County and the rest of the country found out, clear-cuts could not be the only method of forest management. If everything became clear-cut then the industry would have to wait many years for the next generation to reach full maturity. Therefore, the practice of a rotating clear-cut was needed to support the timber industry. Different tracts of land had different aged stands of loblollies so that there would be a continual timber crop. Environmentally it is not a good method because it decreases the diversity of the stand because all the trees will be of the same age and a healthy forest needs a mixed forest of trees of varying ages.

In a study done by Wayne C. Zipperer of the USDA Forest Service, deforestation in Wicomico County caused by clear-cutting creates patterns in the forest cover and affects the diversity of species. According to Zipperer's study, there are five distinguishable patterns of deforestation: internal, indentation, cropping, fragmentation, removal. Each one of these patterns has a direct impact on the habitat quality of the forest patches. The interior, or the middle, of forests is important to many species for protective nesting grounds and refuge.

Between the years of 1973 and 1981 the forest interior declined in Wicomico County by 3.2 square miles. This means species of birds and other animals have been forced to find other forest interiors to nest and seek refuge, thus lowering the diversity in this county. One bird that has been affected is the Red-Eyed Vireo, a Neo-tropical bird that breeds and nests in the mid-latitudes (Wicomico) and spends its winters in the Amazon basin of South America. In order to breed it needs deep groves of shade-trees.

Wildlife also needs connected forests and habitat fragmentation is the most serious threat to having a fully biological diverse forest. The pattern of indentation is the most prevalent pattern in Wicomico County. Indentation occurs when clear-cuts are intruding into the forest interiors creating peninsulas of the forest cover. This practice promotes the loss of interior habitat and promotes forest edge habitat. Forest edge habitat is mostly used as a resting point for migrating creatures and a hunting ground for predators giving the prey less cover in which to hide.

Forests Were First Managed

For thousands of years Native Americans have been managing forests in the Northeastern United States including Wicomico County and the rest of the Eastern Shore. According to Stephen Pyne, a fire historian, Native Americans in the northeast would set fires off in forests in order to make hunting easier. According to the Native American tribes of the eastern United States fire was known as Our Grandfather Fire. Native American "economies were dependent on fire" and without fire their "economies would have collapsed".

Besides burning forests to clear the way for hunting, Native Americans also used fire in the forests to obtain firewood. Although forests provided the first Europeans with timber fuel and game their views of the forest were "an obstacle to agriculture". This European ideology of the forest led to widespread deforestation during the 18th Century. While trees were being cut down to make room for farm fields, the timber cut was being sent down to the islands of Barbados and Antigua who no longer had enough timber because they used up all of their resources.

During these times someone would go out into the forest with an ox-drawn cart and then load the timber onto that. Forests were slashed and burned, meaning people would girdle a tree (removing its bark) in order for the tree to die and dry. After many months once the large trees were dead and dry fire was used to clear large portions of the forests. There is actually a small town in Worcester County called Girdletree which took its name from this slash and burn technique of clearing the forests. According to Jack Wennersten an Environmental Historian, early "Chesapeake farmers and planters had little use for the forest as an aesthetic end in itself. Trees on the horizon irritated their eyes and they wanted to see bare ground".

The forest management techniques of the Native Americans and European settlers were different. The Natives used the forest to benefit themselves however left a minimum impact on the surrounding environment. The settlers viewed the forest as a both an obstacle and resource. Their views exploited the forest to fit their needs and left major impacts to the surrounding environment. Clear-cuts are also dangerous environmentally through nutrient run-off. When clear-cuts take place around bodies of water, excess nutrients in the soil run-off by erosion and into the water. The nutrients promote extra unwanted growth of aquatic vegetation that clouds the water and chokes out fish. The nutrients are allowed to run-off because the root systems of the trees that once held the soil together are no longer present and cannot prevent erosion. The extra nutrients in the water mean excess grow in aquatic vegetation that chokes out fish and other aquatic organisms.

Herbicides

With recent advances in herbicides and fertilizers forest management has turned to some fairly newer practices. Thinning and clear-cutting are still the most widely used management practices however; the use of herbicides has taken over some of the duties of prescribed burns.

Although prescribed burns are still useful to eliminate forest fire fuel on the forest floor they are not needed to eliminate unwanted brush and trees. Herbicides have been used in Wicomico Country in order to allow purely loblolly pines to grow on certain areas and not allow other vegetation. In 1984 the herbicide Imazapyr, commonly know as Arsenal, was registered in the United States and in 1986 was first used in Wicomico County on pine stands after a clear-cut to promote the loblolly pine.

The first application method used with Arsenal in Wicomico County was when a tank of Arsenal was dragged around a clear cut by a skidder in an arc fashion; however this method did not last because it was not cost effective. According to Metzger, since the skidder was too expensive, local foresters contracted helicopter companies to spray sites; this method was more time efficient and less expensive.

Arsenal is applied to clear-cut areas because it stops the growth of unwanted vegetation without harming the commercially important loblolly pine. Herbicides were introduced into the United States in the mid 20th Century and their first uses were for agricultural fields. Prior to Arsenal, the herbicides Velpar and Tordon were used in forestry. Metzger stated that these two herbicides were not always consistent. Tordon and Velpar were not consistent because if it rained after an application it would run-off the desired tract. According to Anthony F. Maciorowski, Branch Chief of Ecological Effects Branch Environmental Fate and Effects Division, "Due to the extreme phytotoxicity, its (Tordon) persistence under typical environmental conditions, and its extreme propensity to leach into groundwater in all soil types, the EEB is strongly recommending against the reregistration of all active ingredients of Picloram. This conclusion is based on the extreme exceedance of the acute levels of concern for non-endangered and endangered terrestrial plants."

Arsenal has been used because it does not run-off if it rains soon after an application and it consistently kills unwanted vegetation. It has both an effect on the diversity of trees and vegetation in the forest and harms forest animals. Arsenal is considered as a non-selective broad-spectrum systematic herbicide that attacks plants stems and root systems. This means Arsenal does not differentiate which plants it kills, however it does not have any effect on the loblolly.

This means endangered and threatened plant species are killed if they come in contact with Arsenal. Maryland currently has ten plants listed as threatened or endangered by the United States Fish and Wildlife Service. These plants include four threatened species: Seabeach Amaranth, Joint-Vetch, Swamp Pink, and the Small Whorled Pogonia. The six endangered species include: The Northeastern Bulrush, American Chaffseed, Smooth Coneflower, Canby's Dropwort, Sandplain Gerardia, and the Harperella. The Canby's Dropwort endangerment is directly due to pine plantations. Canby's Dropwort is native to Wicomico County and was once abundant in the area. It first became threatened when its needed wetland habitat was being lost to the construction of pine plantations. Bulldozers were used to fill in Wicomico's low-lying wetlands and converted them into working pine plantations. Since the Canby's Dropwort needs a wetland habitat in order to survive, the destruction of Wicomico's low-lying wetlands are the reason why this plant is now endangered. Now there is an endangered plant where the application of Arsenal on clear-cut sites can make it even more endangered.

Local foresters claim that the amount of Arsenal required to kill a mammal is rather large. This statement is true, large quantities of Arsenal exposed through animals by oral ingestion, exposure through skin or inhalation is needed for an animal to die. However, animals never take in enough of Arsenal to kill them. Thus it is deemed safe to wildlife, which is false. Just because death is not directly related to the herbicide does not mean it does not have other negative effects on the local wildlife. According to tests performed by Caroline Cox of the *Journal of Pesticide Reform*, Arsenal's acute toxicity has been observed as the cause of many problems with animals. Arsenal causes bleeding and congested lungs in rabbits (the tested animal); congestion was also found in the kidneys, liver and intestines of rabbits as well. Arsenal is also corrosive to the eyes and can cause irreversible damage. When Arsenal is exposed to the skin of animals it has been observed to cause reddening, scaling and crusting and also cause stomach ulcers and intestinal lesions on rabbits. Arsenal also stays persistent in the soil between 60 to 436 days depending on its application and run-off. The foresters are correct, Arsenal does not directly kill animals, however Arsenal does harm animals in a way that may lead to death in another manner.

Thinning

Often, before a clear-cut harvest another technique in forestry management is used to lessen the competition for the loblollies. According to some experts, thinning is the single most important management practice a pine plantation could receive.

The goal in a clear-cut it to get the maximum income from a particular stand in one cut, however, this is not the primary goal of thinning. The goal of thinning is to get the softwood loblolly pine stands ready to provide high value timber in the future.

Thinning is when certain trees are removed in order to encourage the growth of future trees by having them somewhat evenly distributed throughout the stand. In some cases the trees that are removed could be still used commercially and thus the thinning is known as a commercial thinning. In those cases where there is not a market for the removed trees the thinning is known as a precommercial thinning. Thinning is used because if a wooded lot is going to be used commercially the amount of trees on that land needs to be regulated. If there are too many trees growing on the lot it could affect the growth of the rest of the lot produce less desirable timber.

By thinning trees out, it benefits the remaining trees' growth rate, strength and market value. There are three distinct benefits to the tree farmer through thinning. The growth is concentrated on fewer trees allowing them to reach maturity faster meaning more value for the trees that are left. The low value timber does not take up unwanted room allowing only high value timber to continue to grow. Trees that would have died before the final harvest can be marketable and worth money. Thinning is a process that would take place before a final harvest which is often times a clear-cut. An unthinned forest has benefits for some timber manufacturers as well. When the loblolly is cut to be made into poles denser wood is more important and thinning would not be used to keep the trees tall and dense.

On a biological stance, thinning has great economic benefits for the tree farmer as well. The loblolly trees, along with other trees, are in a constant battle between each other for vital nutrients, sunlight, and water to survive. If the stand was just left alone without thinning, the trees would not grow as well because they have to battle for the water and nutrients. With a thinned forest, competition for the nutrients, sunlight and water is much less allowing each tree a better chance at growing into a mature and valuable tree. If natural thinning occurs most trees are too crowded and do not receive enough sunlight, and they just die. According to a North Carolina State University study, a stand that would have had 600 to 1,000 trees could be reduced only to a few hundred by the time maturity is reached at age forty. This displays that if a forest is thinned there would be less competition and more desired trees reach full maturity. One negative effect of thinning economically is that it is more expensive then clear-cutting. Since thinning is more expense the farmer would receive less money rather than if he just had a clear cut. However, with clear-cutting a farmer needs to wait 50-60 years before receiving any kind of money; with thinning some thinned trees are worth money and would give the farmer a money source between final harvests.

The management practice of thinning has been practiced for a long time in Wicomico's timber industry. During the early part of the 20th century state experts in forestry would come to Wicomico County to have lectures to teach farmers and lot owners how to manage their loblolly pine stands so that they could get the most value for their timber. In November of 1924 Fred B. Trenk, an Extension Forester of the University Maryland, gave demonstrations on thinning to local farmers. On April 22, 1926 Dr. F. B. Arenk of the University of Maryland gave a lecture on forestry and handling woodlots to local woodlot owners at the Salisbury Court House. During this lecture Dr. Arenk discussed the shortest possible time in which a loblolly's could be grown and what could be done to help the process. During this same period of time there were also demonstrations given by foresters to farmers on how to thin their forests. These demonstrations and lectures made an impact in the Wicomico area because it gave people economic benefits to their wooded lots. It was also important because it puts more land into active timber production, meaning the dominate forestry management was commercially related.

Prior to the 1980's thinning was an expensive and time consuming practice because it was done manually. Ron Metzger, Wicomico County forester for the State of Maryland, said the biggest change in thinning since he arrived here in the 1980's is the advancement in technology. Thinning is now done with specialized mechanical equipment that cuts down manual labour expenses and also speeds up the process. Now since thinning is a quicker process, more tracts can get thinned in a shorter amount of time creating more productive timber lands. This practice does damage to forest diversity because any competing trees or vegetation are thinned out to allow excess growth to the loblolly pine. Thinning promotes the growth of one species of tree at an even age. A forest with an even-aged single tree specie promotes a balanced habitat to very few organisms.

DOCUMENTATION NATURE OF FOREST MANAGEMENT

There is little documentation on nature of forest management in medieval period inscriptions, though the inferences are that the *Mughal*-rule (1400 A.D.-) saw forest resources as *pleasure* objects. Extensive hunting (for sport) is recorded during the period. Although rulers, particularly Jehangir and Akbar, were fond of roadside trees and gardens, they are argued to have shown little interest in forest conservation, and extensive damage was done during wars with other ruling states.

The reason to sketch out the countenance of ancient and medieval forestry is to emphasize that forest practices could have been as harmful and imperialist in approach as in British-ruled India. More specifically, it seeks to indicate that the state, once finding a forest resource of value, has, through time, tried to exercise control over it. Secondly, as different groups contest for (albeit with different incentives) control, a hierarchy comes to exist both in use and in perception of forests. Hence, where state, for example, could pursue territorial and economic interests in forest terrains, the forest citizen utilized their immediacy to evolve their own set of practices to secure forest product for subsistence and household economy. The various 'niches' in this hierarchy is not always exclusive of each other, and indeed has resulted in the oft-cited contests between the state and the forest citizens. The meaning of participation, hence, would best be found in use of forest in manners that can serve to fulfill needs of a multiple set of users situated at various levels.

Colonial Forestry (1800 – 1947) in British Administration

The newly established British administration in India was initially not alive to the need for careful husbanding of forest resources, and was under the impression that the forest wealth of India was inexhaustible. The British themselves were new to the ideas of scientific forestry, and had no developed forest organization in Britain. The first step in British Indian Forestry came in South India.

In 1800, a commission was appointed to enquire into availability of teak in the Malabar Forests. In 1805, a Forest Committee was constituted to investigate the capacity of forests and the status of proprietary rights over them. Then in 1806, teak (Tectona *grandis*) was reserved as a royal right in parts of south India. Later, in 1855, Governor-general Dalhousie promulgated for the first time an outline for forest conservancy for the whole country called the 'Charter of the Indian Forests'. Following this, Deitrich Brandis was appointed Superintendent of Forests in India in 1856. The Forest Department under him proceeded to transform the working of India's forests, from the initial practice of exploiting them to obtain supplies of timber, to treating them as a growing biological entity of much value, and handling them in accordance with the principles of scientific forestry.

In 1857, after the Indian mutiny, India came under the direct rule of Britain. It was a turning point not only for the civil administration of the sub-continent but also for forest regimes. India was now to be ruled not only for trade interests, but also was taken over by a nation, which was obliged to have a greater control over the territories of her colony. In his letter, dated 1st November 1864 to the Secretary of State for India, the Governor-General pointed out that the idea of allowing individual proprietary rights in forests must be abolished, as such rights might lead to the destruction of forests. However, local governments were allowed to control forest management and were to be only given with policy guidelines by the Government of India. The first major Forest Act came into being in 1865. Under this Act, local governments were empowered to draft rules for law-enforcement in their respective regions. A revised Forest Act in 1878 provided for constitution of Reserved and Protected category forests. During this period (1880-1900) forest *settlemen*t, demarcation and survey of forest land were actively in progress in various provinces. Later, following Dr. Voelcker's report (1897), the Government of India declared in its forest policy, that permanent cultivation should come before forestry, that the satisfaction of the needs of the local population at non-competitive rates, if not free, should over-ride all considerations of revenue, and that after the fulfillment of the above conditions, the realization of maximum revenue should be the guiding factor. Rapid progress was made in organized forestry in the years between 1925-47. There was a long history of working plans and forest-research spanning more than eighty years of 'scientific' management of Indian forests by the time India gained independence in 1947.

FOREST MANAGEMENT SINCE INDEPENDENCE

The Indian Forest Act of 1927, which is modeled on the earlier act of 1878, still defines the legal framework for forest management in the country. Current issues of biodiversity, equity and customary rights do not find prominence in the legislation. With the abolition of *Zamindari* in 1951 large tracts of private forest land were vested with the State, making the Forest Department, as Guha says, the "biggest landlord in the country". The National Forest Policy of 1952 was strongly hinged on scientific forest management which drew largely from existing forestry practices in USSR and the USA. The practice was to encourage maximization of forest revenues and to give priority to agriculture and industry. The second five-year plan (1952-57), which had followed the National Forest Policy, however, also directed the States to aim to put 33 per cent of their land under forest cover.

THE REALM OF NATURAL RESOURCE MANAGEMENT

A series of new incentives and power-relations have been created because of participatory policies in the realm of natural resource management.

Encroachment of forest land, for example, for agriculture and less so for homestead purposes is not uncommon. Understandably, while many of the encroachments have existed before JFM got implaced, with the current management practice, the Forest Department finds it difficult to wrest the land out of these villagers. The encroachment is mainly by people who have leverage within the village community, and are often active members of committee.

Finally, in more extreme cases, defaulters have used the insularity derived from powers and legitimacy of the new committee, to sell timber for commercial purposes. The participatory schemes have somewhat also resulted in limiting both the spheres of influence and activities of the forest department. The policing, for example, is now restricted to 'non-negotiated' territories, areas that are not put under a village's jurisdiction: roads, forest-corridors, reserved-category forests and *haat*, the local-markets (where forest-usufruct is bartered). While in Bihar, the petty officers still patrol forest in uniforms, they enter villages only on being reported or in instances of outright felling of timber. The villagers enjoy a situation where they can, by showing accountability to community-forums, get away with practices that are not allowed within the JFM framework.

Another factor which is common in both the states is the initiation of protection of forest by communities well before the state sponsored participatory policies took shape (often around 1985, encouraged by Social Forestry). The villagers vividly remember the hardships they have faced on many occasions due to unrestrained felling (often by the State through planned coupe-felling), and there is a constant reminder, that unless some manner of restraint is kept, forests will then turn yet again into Open Access Resource. While the State do not provide the FPCs with any regulatory powers, many of the norms have been imposed through practice such as extracting fines from the defaulters.

INDIGENOUS FORESTRY: PERCEPTIONS AND METHODS

The working methods employed in informal activities and transactions are not hard to ascertain. Since the onset of participatory schemes, there has been decrease in policing and violent confrontations between villagers and forest officials. This marked *withdrawal* of forest department staff from actual resource area has prompted many inherent, albeit veiled, activities to come to the fore. JFM-led village-forums often 'regulate' these practices, without adhering to the boundaries (marking out concessions and rights) that the management agreement has prescribed.

Creation of 'Zones of Exclusion'

A key way the FPCs have laboured to protect local forest, is through creation of 'zones of exclusion'. The forest that lie under a village's jurisdiction are marked 'out of bounds' for people outside their community, and, if violated, fights are taken up with infiltrators from neighbouring villages. Again, the boundary that constitute these zones is not always that have been delineated by the FD. In Bihar, for example, the villagers, depending on their perception, mark out territories in forest-land, as their *rakhat*, or an area that is under protection, to exclude outsiders.

Within the community setting, however, runs a noteworthy egalitarianism: not only they maintain a understanding of a universal need of extracting fuelwood (usually of fallen twigs, branches, of green shrubs, and at times of live trees, branches) but there also exists a consent, in principle at least, for extraction of timber for meeting essential needs like construction and repair of houses and for agriculture. These activities generally need a 'clearance' within the forum or carry understanding of the local leader. In Bhagwati Chowk, for example, it suffices to have the understanding of Poltu Singh (though many others would be informed). In Ober, the NGO-run club authorizes extraction of trees for a small donation.

The villagers own reading into Joint Forest Management systems is, at best, vague, and their activities, in the assumed participatory atmosphere, are instead informed through their own interpretations. They, though, are aware of the committees' responsibilities to protect forests (chiefly, from 'others') through guard-duties (and self-restrain), the nature of rights they are to have in return is uncertain. It is commonly understood that forest is not open to indiscriminate logging, but would, however, provide for the fuel and timber to cover basic needs. The practices are grounded on strong notions that forest belongs to the entire village community, and while rights on its land-tenures rest with the government, the villagers can not be alienated from their day to day use, 'the *sircar*, after all, can not take forests away', the villages reason. In their interaction with the Forest Department, the villagers frequently report of disappointment. They complain of not being rewarded in instances when they reported defaulters to the department, nor have they received material-support to aid them guard forest. Importantly, Forest Department seen as a part of the 'government', is expected to bring in development projects, while this is seldom the case. Another requirement of the JFM agreement, that villagers would replace the Forest Department with provision of guarding duties, flagged after initial bouts of enthusiasm.

Finally, the principal way JFM departed from previous participatory initiatives, by promising cash returns to FPC members, has produced mixed results. In Bhagwati Chowk with 37 households and nearly 150 hectares of forest under the FPC's jurisdiction, the two annual harvests that were carried out by the FD have fetched each of the households with significant amount of cash. This has rarely been the case in other situations. Amratoli, the neigbouring FPC with 240 hectares of forest and 246 households has a much lower forest/

household ratio and cash return on harvest is hardly an incentive for them to institutionalize JFM practices. Bihar, on the other hand has yet to carry out its first harvest under JFM agreement. Here, and in instances where the cash-incentive is not sufficient, the reason for involvement lies more in the actual control of the resource and also in access, this newly found base provides, to stronger institutions.

Gender Divide in Community Forestry

Division of labour in extraction of forest usufruct is sharp. Majority of fuelwood extraction is carried out, often on a daily basis, by women members of the household. Forest in some ways is their exclusive work-zone, so much so that men find it socially awkward to scrounge forest-floors. The role inter-changes when men have to cut trees for timber, for construction of houses and agricultural implements. Despite woman foraying into forest more frequently, the control that men exercise over forestry practices is unmistakable. Men think they can always stop women if they go *haywire*. A usual explanation by women, for extraction of wood, is, '*pooch kar late hain*, we ask before we bring forest usufruct'. The case changes slightly in Bengal, where effeminated men do, at times, collect fuelwood and even barter them in markets in the nearby township. Despite a less profound role in forest usufruct collection, men (as, also, shown in her thesis by Sarah Jewitt, and quite in contradiction to Vandana Shiva's eco-feminist principles) demonstrate considerable knowledge of time, species and pathways that regulate choices, in forest, on part of their womenfolk. While forest provides women with a *zenana* womb, where they socialize with other female members of community, it remains an 'alien' quarter where familiarity is seldom past a point, never beyond a time: what with herds of elephants, and one respondent even saying 'a small tiger' that are present in forests. Women rarely venture out on their own.

INDIGENOUS RIGHTS AND ACCESS TO FORESTS

It is apparent that in the nineteenth century and right up to the 1970s, indigenous rights and access to forests were contested by state agencies claiming to act in the national interest. Many forest-dependent communities, tribal (*adivasi*) groups in particular, were stigmatized as forest-destroying communities. In the face of these contests (and rapid degradation of forest cover) there has been growing support in India for the idea of placing some 'power' back in the hands of the rural poor. Since the late 1970s a number of social forestry programmes have been developed and the emphasis of state forest policies has shifted from commercial forestry to that of meeting needs of the forest dependent population on a priority basis. The most recent of these programmes (1988 -) has embraced a philosophy of Joint Forest Management (JFM).

Social Forestry Development

Following the Agricultural Commissions' recommendations in 1976, forests that had remained under total control of the states became a *concurrent* subject. The government in New Delhi was to play central role in formulating policy guidelines and in coordinating forestry programmes. As the interests of the centre and the states do not necessarily coincide, the government in New Delhi could arguably take a long-term view. This arrangement, allows conservationists, international agencies and voluntary groups, despite their low numerical strength, to have considerable influence on the direction of policy at the level of the centre. Also, state governments are subject to regional and local pressures, which may not be a factor for decision making at the centre, it was argued.

Raise Forest on Private Land

In the 1970s members of the public were invited to raise forest on private land. By that time it was widely recognized that destruction of forests would not cease without obtaining participation from forest dependent communities. The nation had by then geared to address the subsistence and cultural dependence of its people. The period coincided with the social justice call of Prime minister, Indira Gandhi (5th five year plan). Reflecting the national rhetoric of social justice and equity, rural communities were to find increased (official) access to forest usufruct. Thus a shift, from pure regimes to shared regimes took place. India as a leader in social forestry) took a series of participatory interventions: village woodlots, strip plantation (roadsides, alongside railways), farm forestry, agro-forestry, home gardens, urban forestry, fuelwood plantations, festival of trees (*van mahotsava*), education camps, nurseries, and research and training.

However, a clear disjunction between the programmes' intended goals and actual products emerged. While the targets of farm forestry (in many states), for example, were considerably over-achieved, those from community forestry were less successful. It were larger farmers who primarily took up farm forestry and adopted commercially viable species, such as eucalyptus, for plantation. Villagers invariably had an incentive in income generation. Also, the project was alleged to have encouraged little participation from tribal population and villagers were reluctant to make long-term investments (Shepherd, 1985). Vandana Shiva, the eco-feminist writer, has been vociferous against the programmes which she thought promoted mono-culture, did not encourage involvement from women and from which particularly poor seasonal migrant labourers lost employment as tree planting is not a labour intensive occupation.

Afforestation of wasteland to her, hence, was 'privatization' of the commons. Social forestry found willing funding bodies in World Bank, USAID, DANIDA, ODA, SIDA, EEC, OECF. These donors progressively attempted to influence the direction of policy especially in regeneration of degraded forest areas and for wildlife conservation. The international agencies came under attack for promoting a 'top-down' approach which the participatory programme was to do away with. They were alleged to have vested interest in country's corridors of power. Also, the forest bureaucracy was criticized for its inability to shake off power structures that are conditioned to emphasize policing, protection, and state production. Also, the department represented a long history of protecting the interests of forest-dependent industries. These factors despite policies at the centre to give priority to fuelwood and subsistence needs of the village community, failed to marginalize the private commercial interest in the forests.

Joint Forest Management: The Current Paradigm

Notwithstanding the criticisms and failures that social forestry faced, the programmes succeeded in initiating the processes for participatory forest regimes, devolution of decisions and recognition of villagers' subsistence needs. In 1990, the Ministry of Forestry and Environment, in New Delhi advised the state governments to take up Joint Forest Management (JFM).

Not surprisingly, Joint Forest Management systems could get attention both from the group (like Guha and Gadgil) that wishes to see in JFM, the chance of restoration of moral economy through increased tenurial and usufruct rights for the rural communities and exercise of 'sustainable' indigenous practices, and those (like Rangan) who, on the other hand, are keen to find solutions within existing institutions by way of 'pressuring states to intervene on behalf of marginalized communities, to ensure equitable access to the potential benefits of economic development'. Results from JFM have indeed been encouraging. JFM is reported to be instilling new attitudes and behaviour in forest bureaucracy towards the villagers, and vice versa. Many forest beat officers have reported that their primary incentive to encourage JFM is the vastly improved relations they have come to cultivate with the community. NGOs too have found themselves accommodated in the new regime, as an agent of change and promotion. Through group exercises, discussions, meetings and monitoring they are to instill participatory practices both in members of Forest Protection Committees and the Forest Department officials.

There have been several shifts in management of forests by the state and these, in turn, have been reflected in attitudes adopted by the user groups. The people, in the current setup, have come to view the state, NGOs and grassroot organizations (*panchayats*, for example) as bodies they can use to bring development in their community. The solution to equitable use of forests is not to isolate forest citizens in realm of some imagined sustainable practices but to chart out on what sorts of institutional links and relationships could be established between the rural communities and other institutions to enrich this dialogue. The chapter contends that while many predicaments arising from forest conflicts between state and people have been successfully attempted at, the state not only has to take a more imaginative approach to include the practices that flourish informally at FPC level, but also reach out to intervene in various facets of rural development forestry. The findings here constantly are pointers to the peoples' desire to involve the state (and other institutions) for development and access to market and available funds, as against leaving out the state in such processes.

Research Methodology

The paper compares the States of Bihar and West Bengal, sharing contiguous boundary and similar socio-cultural environment and forest ecology. It examines the characteristics that mark forest activities (extraction, consumption, protection) among forest citizens in Bihar and Bengal. The hypotheses that are tested include that the Forest Protection Committee, a key stakeholder in JFM, will be motivated to invest time and other household resources to meet the *mandatory* demands of JFM (forest-guard duties, monitoring, attendance in meetings, forest works, etc.) over a number of years in anticipation of sharing the usufruct gains associated with afforestation and other forest management schemes; that the age old conflict over establishment of rights over forest resources would be amicably resolved (now that government recognizes the claims of villagers, the villagers would in turn recognize the jurisdiction of forest land boundaries); and finally, that the 'equity' rhetoric would be served. JFM would aid the local community in various village development works, and in increasing their income.

The means we have depended on include literature-review and archival analysis. Theories and archives provide the background to which case-studies are directed, and against which the results from the case-studies have been tested. The research looks into positions taken by ecological institutions in the JFM set up. The field work included interviews, group discussion, attending meetings, participation-observation, observational techniques and secondary data collection (official documents/ progress reports). Interview methods were mostly semi-structured and open-ended. Research carried out in one State was mirrored in another.

ACHIEVE SUSTAINABLE FOREST MANAGEMENT

Historically, Indian forest policies have alienated people from the forests, thereby, exacerbating the rates of deforestation. Post-independence forest policies contributed to an expansion in agricultural production, met industrial demand for raw materials, and tightened control of forest lands through restricted access to forests and forest products. Protection policies increased the hardships of vulnerable social groups by denying them access to forests.

While the state took responsibility for managing forest resources, it did not have the commensurate resources to effectively manage and police the forests from traditional users. Before state intervention, forests were managed as communal property; the crucial role of forests in the economic subsistence of individuals, families and community was the basis for managing them as communal resources. A failure to recognize community control of forests led to a collapse in institutional norms that were instrumental in protecting and managing forest resources for local use. A shift in property rights to the state steadily undermined the rights of tribals to use and extract forest resources.

Involvement of rural communities living close to forests in protection and management of forest resources is enshrined in the National Forest Policy 1988.Translation of policy found expression in the resolution of Government of India, Ministry of Environment and Forests issued in June 1990.It envisaged that in lieu of the participation, the local communities will be entitled to sharing of usufructs in a manner specified by the concerned State Forest Departments. This led to the initiation of Joint Forest Management (JFM) programme. Importance of the programme is evident from the fact that the Government of India has constituted a "JFM Network" with the Inspector General of Forests, Government of India as the Chairman. The objectives of the network are (i) to act as a regular mechanism of consultation between various agencies engaged in JFM work in the country and (ii)To obtain constant feedback from various stakeholders on the JFM programme for proper policy formulation and suitable direction to States.

World leaders adopted the Millennium Declaration at the Millennium Summit in September 2000. The proportion of land area covered by forest globally is one of the indicators for the seventh MDG i.e to ensure environmental sustainability.In addition to quantitative, time-bound targets, the Millennium Declaration calls for other actions, including intensified efforts for "the management, conservation and sustainable development of all types of forests", an international commitment to sustainable forest management made in 1992 at the United Nations Conference on Environment and Development (UNCED) and embodied in the Forest Principles. Subsequent intergovernmental deliberations to promote progress towards sustainable forest management took place in the Intergovernmental Panel on Forests (IPF) and Intergovernmental Forum on Forests (IFF) from 1995 to 2000, and continue in the United Nations Forum on Forests (UNFF) as well as in other fora.

SOCIAL FORESTRY IN INDIA

Social forestry means the management and protection of forests and afforestation on barren lands with the purpose of helping in the environmental, social and rural development. The term, social forestry, was first used in India in 1976 by The National Commission on Agriculture, Government of India. It was then that India embarked upon a social forestry project with the aim of taking the pressure off currently existing forests by planting trees on all *unused and fallow land.*

SOCIAL FORESTRY PROGRAMME

Government forest areas that are close to human settlement and have been degraded over the years due to human activities needed to be afforested. Trees were to be planted in and around agricultural fields. Plantation of trees along railway lines and roadsides, and river and canal banks were carried out. They were planted in village common land, government wasteland, andPanchayat land.

Involvement of Common People

Social forestry also aims at raising plantations by the common man so as to meet the growing demand for timber, fuel wood, fodder, etc., thereby reducing the pressure on the traditional forest area. This concept of village forests to meet the needs of the rural people is not new. It has existed through the centuries all over the country but it was now given a new character.

With the introduction of this scheme the government formally recognised the local communities' rights to forest resources, and is now encouraging rural participation in the management of natural resources. Through the social forestry scheme, the government has involved community participation, as part of a drive towards afforestation, and rehabilitating the degraded forest and common lands.

Need of Social Forestry

This need for a social forestry scheme was felt as India has a dominant rural population that still depends largely on fuelwood and other biomass for their cooking and heating. This demand for fuel wood will not come down but the area under forest will reduce further due to the growing population and increasing human activities. Yet the government managed the projects for five years then gave them over to the village panchayats (village council) to manage for themselves and generate products or revenue as they saw fit.

Types

Social forestry scheme can be categorized into groups; farm forestry, community forestry, extension forestry and agroforestry.

Farm Forestry

At present in almost all the countries where social forestry programmes have been taken up, both commercial and non commercial farm forestry is being promoted in one form or the other. Individual farmers are being encouraged to plant trees on their own farmland to meet the domestic needs of the family. In many areas this tradition of growing trees on the farmland already exists. Non-commercial farm forestry is the main thrust of most of the social forestry projects in the country today. It is not always necessary that the farmer grows trees for fuelwood, but very often they are interested in growing trees without any economic motive. They may want it to provide shade for the agricultural crops; as wind shelters; soil conservation or to use wasteland. Farm Forestry is another name for Agroforestry; a part of Social Forestry.

Community Forestry

Another scheme taken up under the social forestry programme, is the raising of trees on community land and not on private land as in farm forestry. All these programmes aim to provide for the entire community and not for any individual.

The government has the responsibility of providing seedlings, fertilizer but the community has to take responsibility of protecting the trees. Some communities manage the plantations sensibly and in a sustainable manner so that the village continues to benefit. Some others took advantage and sold the timber for a short-term individual profit. Common land being everyone's land is very easy to exploit. Over the last 20 years, large-scale planting of Eucalyptus, as a fast growing exotic, has occurred in India, making it a part of the drive to reforest the subcontinent, and create an adequate supply of timber for rural communities under the augur of 'social forestry'.

Extension Forestry

Planting of trees on the sides of roads, canals and railways, along with planting on wastelands is known as 'extension' forestry, increasing the boundaries of forests. Under this project there has been creation of wood lots in the village common lands, government wastelands and Panchayat lands.

Schemes for afforesting the degraded government forests that are close to villages are being carried out all over the country.

Agroforestry

In agroforestry, silvicultural practices are combined with agricultural crops like leguminous crop, along with orchard farming and live stock ranching on the same piece of land. In lay man language agroforestry could be understood as growing of forest tree along with agriculture crop on the same piece of land.

In a more scientific way agroforestry may be defined as a sustainable land use system that maintains or increases the total yield by combing food crop together with forest tree and live stock ranching on the same unit of land, using management practices that takes care of the social and culture characteristic of the local people and the economic and ecological condition of the local area.

FORESTS AND THE MILLENNIUM DEVELOPMENT GOALS

The MDGs call for the integration of the principles of sustainable development into environmental policies. Environmental sustainability is being mainstreamed in forest policies around the world, particularly since UNCED, while the integration of the goals of poverty and hunger reduction in forest policies and plans is less widespread.

Community-based forestry, or participatory forestry, is particularly well placed to address poverty reduction. Community-based forestry is now well accepted and established in various countries in all regions, and programmes are beginning to generate financial and other benefits. Improving local peoples' rights and access to forest resources is a tool to the success of community-based forestry programmes. However, much still remains to be done to clarify and secure access rights. Many countries are working to strengthen forest governance, some through decentralization processes that allow the poor to derive more benefits from forests and be more involved in decision-making and forest management itself.

Although improving rights and access to forest resources and developing small-holder forest-based enterprises (including through community-private sector partnerships) show particular promise for poverty reduction, local political and economic realities, opportunity costs for the use of local resources, and other factors may prevent the poor from benefiting from community-based forestry programmes to the extent intended.

Intersectoral coordination is important for the achievement of all MDGs, but is particularly critical for reducing poverty and hunger and ensuring environmental sustainability, which are highly cross-sectoral by nature. Improved intersectoral cooperation and coordination will help efforts both to integrate the principles of sustainable development into forest-related policies and to integrate forests into sustainable development plans.

Joint Forest Management

Forest-based poverty reduction efforts tend to be linked to other land uses and should form a part of rural development strategies. Conversely, the potential for forests and trees outside forests to contribute to environmental sustainability cannot be fully realized without intersectoral cooperation and coordination. Intersectoral coordination, although difficult and time consuming, is necessary for sound decisions on land use and resource allocation, particularly when there are trade-offs between national development goals.

National Forest Policy in India treats forests as environmental and social resource. With the initiative of assigning ownership of Non Timber forest Produce (NTFP) to the local communities including the grass root level democratic institution for enhancing their livelihood opportunities and also improving their income with the value addition.

India has shifted the approach of forest management from regulatory to participatory mode of management with the resolution promulgated in 1990. At present, more than 17 million forests is managed by almost 10,000 Joint Forest Management Committees with the benefit sharing mechanism. In addition, the Government of India is in process to frame legislation for the settlement of tenurial rights of the forest dwelling communities mainly tribal on forests. This would definitely help in reducing the poverty of forest dwelling communities. The JFM resolution was circulated by Ministry of Environment and Forest in the year 1990 and 2000. JFM is a government resolution. A government resolution is a executive order or opinion of the legislature. A resolution does not have any legal backing.JFM as the term indicates is the management of forest by more than one party. In India there are two parties: the government represented by the Forest Department and the people living in villages located within forest or on the fringes.

There are two major reasons behind introducing JFM: one that the government's management system was not succeeding in arresting growth of forest degradation and deforestation.Second a new management paradigm was evolving in which the local people's participation was found to be an appropriate and promising tool in arresting forest degradation.

However in pre-independent India the concept of JFM didn't exist. The first National Forest Policy was adopted in 1894. Following were the guide lines

- Ensure maintaneance of adequate forest cover
- Meet the needs of local people.
- Collect maximum revenue after meeting the needs of the local people.
- Give priority to permanent cultivation over forestry land

In post independent India there was a shift in policy. In1988, the new forest policy was adopted which covered all the sustainable management approaches. The new policy had a few unique features. Which were as follows;

- Maintenance of environmental stability and restoration of ecological balance, soil and water conservation
- Conservation of natural heritage and genetic resources.
- Increasing productivity to meet the local needs then the national need
- Creating massive peoples participation movement to protect forest and tree cover and achieve the objective of reducing pressure on existing forests and meeting peoples need.
- Deriving economic benefits must be subordinated to these principal aims.

This initiated a process of reform at the local policy and operational level of forest management ensuring that the Forest Department developed close collaboration for protection and sustainable management of forests.

The aim was –Involvement of village communities and voluntary agencies of degraded forest land.

Important guidelines were as follows:

- The programme should be implemented under an arrangement between a voluntary agency or beneficiaries and the State Department.
- No ownership rights or lease should be given over the forest land.
- The beneficiaries should be entitled to share usufructs to the extent and subject to conditions prescribed by the State Government.
- Access to forest land usufructs should be available only to benefactress who get organized into a village institution especially for forest regeneration and protection. This could be through a village panchayat or a Village Forest Committee.
- The beneficiaries should be given usufructs like grass, lops and tops of branches and minor forest produce. If they successfully protect the forest they will be a portion from the sale proceeds when they mature.
- Areas selected from the programme should be free from claims from any person who is both a beneficiary under the scheme.
- The selected site should be worked in accordance of Working Scheme duly approved by the state government.Such a scheme may remain in operation for ten years and revised after that. The working scheme is prepared in consultation of with the beneficiaries
- It should ensure that there is no grazing at al on the forest land protected by the Village Forest Committee. Permission to cut and carry grass free of cost should be given so that stall feeding is promoted.
- No agriculture should be promoted on the forest land.
- Cutting of tress should not be permitted before they are ripe for harvesting. The Forest Department should not be permitted to cut to cut trees protected by the Village Forest Committee except in a manner prescribed in the working scheme. In case of emergency needs the village community should be taken into confidence.
- The Forest Department should closely supervise the work.If beneficiaries are unable to perform their assigned duties in a satisfactory manner the usufctory benefits will be withdrawn without giving any compensation.

Such set-up, however suffered from certain flaws:

- Bye- laws have not been formulated for the functioning of the JFMC though now most of the states have issued executive orders for the functioning of JFMC but these executive orders are not binding on JFMC. The aim was to decentralize the process and make JFMC.
- Minor Forest Produce (MFP) has not been defined neither by the State legislature or by the Centre.

- Central and State/UT Governments have issued guidelines for the creation and functioning of JFMC but these guidelines are not in conformity with the provisions of the Constitution
- Through the Constitution (Seventy-third Amendment) Act, 1992 and Panchayat (Extension to the Scheduled Areas) Act, 1996 ownership rights over minor forest produce (MFP) have been given to Village Panchayats. Now on the same resource base *i.e.* NTFP/MFP we have two sets of groups having ownership and while the Panchayats have a legal backing, JFMC don't have. conservation have been raised while the proponents of Panchayats have termed JFM as a parallel institution against the spirit of the Constitution.
- The JFM resolution only provides 20 per cent of share to be given the Joint Forest Management committees, while rest of the income would go to the Forest Department. There is an obvious unequal distribution of benefit sharing between the parties. These committees would have to protect the forest for ten years,and would only receive 20 per cent of the share.
- The national resolution provides that at least 33 per cent of the seats shall be reserved for women in the specified committees. However states like Rajasthan are not following this provision.

Panchayati Raj System

The concept regarding Panchayati is contained in the Part IX of the 73rd amendment of The Indian Constitution.This amendment came into force in on 24th April 1993. It institutionalized the third stratum of government (Panchayats) at the local level.

Through the Constitution (Seventy-third Amendment) Act, 1992) Part IX "The Panchayats" was inserted in the Constitution which paved the way for "Village Panchayats" by making provisions for the constitution of Panchayats, their composition, election, powers, authority, responsibility, audit, etc. This Act doesn't apply to Schedule Area referred to in clause (1) and tribal areas referred to in clause (2) of Article 244 and certain other specified areas Village Panchayats have been given the responsibility of social/farm forestry, minor forest produce (MFP), and soil conservation through Eleventh Schedule.

"Panchayat" means an institution (by whatever name called) of self-government constituted under article 243B, for the rural areas. The amendment it fails to define what is 'self government'. It does not clarify whether self government means complete autonomously or extension of the state.It is left to states to derive their own interpretation regarding the nature of self government at the local level.By the word "nature" of self government it is meant reservation to be provided for the marginalized section of the society, the powers given to each level of self governing relating to administration, execution and performing the defined roles would be different in each State. Thus this will defeat the purpose of uniformity regarding the nature and concept of self government at an all India level.

The flaw again lies that the amendment has failed to assert the nature of 'Power And Authority'.This task is left to the state legislatures.Again defeating the purpose of uniformity. The consequences will be such that either the states will deny power to the panchyats instead of becoming autonomous local governing units they will basically remain in bureaucratic control.The burecratic trend can be seen most of the states.It is seen in the State of Goa power and functions have not yet been transferred to the Panchayats. As per the law there should be continuous elections every five years.However the statistics says that this status is not achieved in most of the states.None of the states except WestBengal, Tripura and Rajasthan have not been able to hold continuous elections every five years. The judgement provided by the Supreme Court states that it is mandatory for every state to hold elections to Panchayat.

Giving powers to the third stratum of government for effective management of resources at the district,intermediate and village level. Parallel institutions have come about for management of the natural resources which

Participatory Approach

According to United Nations Development Programme's report 67.7 million people belonging to "Scheduled Tribes" in India are generally considered to be 'Adivasis', literally meaning 'indigenous people' or 'original inhabitants', though the term 'Scheduled Tribes' is an administrative term used for purposes of 'administering' certain specific constitutional privileges, protection and benefits for specific sections of peoples considered historically disadvantaged and 'backward'.

Out of the 5653 distinct communities in India, 635 are considered to be 'tribes' or 'Adivasis'. With the ST population making up 8.08 per cent (as of 1991) of the total population of India, it is the nation with the highest concentration of 'indigenous peoples' in the world. 68 million tribals who inhabit forests and wild lands throughout India, Adivasis have evolved an intricate convivial-custodial mode of living.

The Beginning of Land alienation

Introduction of the alien concept of private property began with the Permanent Settlement of the British in 1793 and the establishment of the "Zamindari" system that conferred control over vast territories, including Adivasi territories, to designated feudal lords for the purpose of revenue collection by the British. This drastically commenced the forced restructuring of the relationship of Adivasis to their territories as well as the powe relationship between Adivasis and 'others'. The predominant external caste-based religion sanctioned and practiced a rigid and highly discriminatory hierarchical ordering with a strong cultural mooring.

After the transfer of power, the rulers of the Residency Areas signed the "Deed of Accession" on behalf of the ruled on exchange they were offered privy purse. No deed was however signed with most of the independent Adivasi states. They were assumed to have joined the Union. The government rode rough shod on independent Adivasi nations and they were merged with the Indian Union.

This happened even by means of state violence as in the case of Adivasi uprising in the Nizam's State of Hyderabad and Nagalim. The Constitution of India, which came into existence on 26 January 1950, prohibits discrimination on grounds of religion, race, caste, sex or place of birth and it provides the right to equality, to freedom of religion and to culture and education. STs are supposedly addressed by as many as 209 Articles and 2 special schedules of the Constitution - Articles and special schedules which are protective and paternalistic.

Article 341 and 342 provides for classification of Scheduled Castes (the untouchable lower castes) and STs, while Articles 330, 332 and 334 provides for reservation of seats in Parliament and Assemblies. For purposes of specific focus on the development of STs, the government has adopted a package of programmes, which is administered in specific geographical areas with considerable ST population, and it covers 69 per cent of the tribal population.

9

Land and Farm Resources Management in Forestry

LOCAL CAMPAIGNS TO BOOST FARM FORESTRY

Two Farm forestry Service Groups (FFASGs) were formed in Kooki and Kakuto counties, each comprising five individuals. These groups are responsible for disseminating information on the role of Farm forestry to the communities as well its contribution to biodiversity conservation. They will give advice on nursery bed construction and management, and silvicultural practices, among others. In addition, after contacting and talking to some members working with forestry and community projects, and sponsored by RSGF, we have agreed to form a Rufford Small Grants Beneficiaries network. This network will share knowledge about their works. We will be able to advise each other and learn from each other in terms of successes and difficulties. The aim of the network is to carry out exchange visits with beneficiary farmers, stakeholders and other interested communities to facilitate knowledge sharing. This spirit is hoped to continue even in our next projects. At the moment four individuals have subscribed to this and we hope to bring more on board.

Main partners/collaborators

- Busia Women producers Association (BUWOPA), Uganda
- Integrated Rural Community Development Initiative (IRUCODI): the NGO that I am affiliated to.
- Pro-Biodiversity Conservationists of Uganda (PROBICOU)
- Wildlife and Animal Resources Management Department (WARM), Makerere University, Kampala.

Other Partners/Collaborators

- Natural Enterprise Development (NED), Uganda
- Uganda Women in Development (UWD)

Over the years, we have been approached by some NGOs, and individuals interested in knowing more about the work I have been doing with the help of RSGF. Some were seeking consultancies while others wanted to have an experience of the work we are doing in Rakai. It is important to note that from all the associations, meeting, and phone calls, we as a team have management to establish links with the above bodies. These links are bound to stay active and from them we shall expand on the work we are doing to include a broader community, and further promote conservation issues countrywide, as well as region wide. A few individuals have been enlightened and are compelled to apply for funds to do related work in other areas of the country. Such spirit expressed by young conservationists is very promising.

Introducing Social Farm forestry Networks

Networks have been formed among some beneficiary farmers and other stakeholders in Kabonela ad Kakuto More networks will be formed as he project continues in the next phase of the project so that there is communication amongst all RSGF beneficiary individuals. The Social Farm Forestry Networks have been formed by the help of the newly created FFASG and by the end of the project period in future; it is projected that the entire district will have a lot of knowledge about farm forestry and its roles in biodiversity conservation s well livelihood improvement.

TREE PLANTING

Rolling out tree planting to other areas was based on the shared knowledge values in biodiversity conservation and improving rural livelihoods. Tree nursery construction and planting of seeds was started in late April 2009.

Table. Overall number and species of trees planted

Tree species	Quantity				Comment
	Phase 1	Phase 2	Booster phase	Total	
Eucalyptus spp	2500	2000	3000	7500	Income generation, cheap to raise and rapidly re-grows after being cut down
Pinus spp	0	800	1500	2300	Grow rapidly and straight and good for storm formation
Orange trees	0	800	0	800	Carbon sequestration, and nutrition in form of fruits
Mango trees	0	0	1100	1100	Carbon sequestration, and nutrition in form of fruits
Ficus spp	0	0	1000	1000	Soil improvement tree species and provides good habitat for many birds
Albizia spp	0	0	1000	1000	For holding the soil in place
Acacia spp	300	200	200	700	Nitrogen fixing
Moringa Oleifera	200	200	0	400	Herbal medicine
Total	3000	4000	7800	14800	

The parishes/villages of Kakuto, Kabonela and Kooki were the main stations.Seven thousand eight hundred trees were planted in the booster phase of the project; an excess of 1800 trees from the 6000 that were originally planned for in the project. In total, 14,800 trees have been planted since the commencement of the project.

Meeting with Local Leaders and Provision of More Extension Services

It is a rule of thumb to meet with Local council chairpersons, opinion leaders and key informants before start of any grassroots project. As routine these meetings were carried out and the way forward was drawn. The local leaders were very appreciative of the work and commended RSGF for their immense concern about nature conservation in the sub-Saharan Africa.

Continued Training

Stakeholders were educated about the following and given practical hands on in some areas:

- Silvicultural practices
- Seedbed management
- Fertilisation
- Increasing patchiness
- Increasing microenvironment
- Secondary succession
- Vegetation climax
- General Biodiversity conservation

LAND USE: ENVIRONMENT AND THE SUSTAINABILITY

Land-use changes can affect the environment and the sustainability of production. Because impacts on the environment—including erosion, water quality, and wildlife habitat—are typically not reflected in private profit calculations, land-use choices that are optimal for an individual may not be optimal for society. This difference suggests the possibility of public policies that more closely align land-use decisions with social objectives. The allocation of a fixed land base among competing uses is determined by the relative returns to the different uses, which vary according to land quality and location. A landowner seeking to maximize profits will allocate a land parcel to the use that yields the highest expected economic return, after the costs of conversion. As relative returns change along with market conditions, technological advancements, or government policies, land-use patterns tend to adjust accordingly. Land-use change is dynamic. With the exception of urban land, changes occur to and from major land uses. For example, 44 million acres left the cropland and pasture category from 1992 to 1997 while 21 million acres shifted into the category, resulting in a net loss of 23 million acres.

MAJOR LAND USES

This series contains acreage estimates of major uses by region and State, coinciding with each census of agriculture from 1945 through 2002. Because Alaska and Hawaii have very little crop area, we focus on the contiguous 48 States. The total land area of the 48 contiguous States is approximately 1.9 billion acres, with an additional 365 million acres in Alaska and a little over 4 million acres in Hawai.

Grassland pasture and range, the largest use of land, accounted for 584 million acres (31 per cent) of the 48 States in 2002. This compares with 636 million acres in the mid-1960s. Due to improvements in the forage quality and productivity of grazing lands, less pasture and range is needed to sustain grazing herds.

The inventory of domestic animals, particularly sheep, has also been declining in recent years, further reducing pasture/range demand.

Forest-use Land

Table. Major uses of land, United States, 2002[1]

Land use	48 States Million acres	US	48 States % of total	US
Cropland[2]	441	442	23.3	19.5
Grassland pasture and range	584	587	30.8	25.9
Forest-use land[3]	559	651	29.5	28.8
Special uses[4]	153	297	8.1	13.1
Urban	59	60	3.1	2.6
Miscellaneous other land	97	228	5.1	10.1
Total land area[5]	1,894	2,264	100.0	100.0

Note:

[1]See Major Land Uses for estimates of major uses by region and State, coinciding with each census of agriculture, from 1945 through 2002.

[2]All land in the crop rotation (used for crops, used for pasture, idle cropland). Includes about 34 million acres idled under the Conservation Reserve Programme.

[3]Total forest land as classified by the U.S. Forest Service minus an estimated 98 million acres of forested land used for parks, wildlife areas, and other special uses.

[4]Rural transportation areas, land used primarily for recreation and wildlife purposes, various public installations and facilities, farmsteads, and farm roads/lanes. Excludes urban land in contrast to Major Land Uses, Aggregate Data.

[5]Distributions by major use may not add to totals due to rounding.

Forest-use land, the second largest major use, declined from about 32 per cent of total land in 1945 to about 30 per cent in 2002. A broader category, all land with forest cover, comprised 33 per cent of the land base in 2002.

While forest-use land increased 1 per cent between 1997 and 2002, it declined from 612 million acres in 1964 to 559 million acres in 2002. Much forest-covered land is in "special uses" that prohibit forestry uses such as timber production. Forested land in these special uses increased from 23 million acres in 1945 to about 98 million acres in 2002.

Cropland comprises the third largest use of land, covering 23 per cent of the contiguous States in 2002. Since 1945, cropland ranged from a high of 478 million acres in 1949 to a low of 441 million acres in 2002. Total cropland has trended downward since the late 1960s, and decreased by 13 million acres (3 per cent) from 1997 to 2002. The total cropland base includes cropland used for crops, cropland used for pasture, and cropland idled. These components vary more than total cropland. Since 1945, the amount of cropland used for crops has ranged fromas much as 383 million acres in 1949 and 1982 to a minimum of 331 million acres in 1987. Total acreage used for crops exhibited two major cycles between 1945 and 1987, with cropland moving from idle to crop use and back again. Cropland used for crops increased from 331 to 349 million acres over 1987-97, and then declined to 340 million acres in 2002, about 5 per cent below the average acreage for 1910-97. Since 1945, cropland used for pasture varied from 47 million acres in 1945 to 88 million acres in 1969.

LAND ALIENATION AND PAUPERISATION

The colonial state imposed the land laws based on this worldview on the traditional communities. This imposition affected the dominant castes as well as the tribal communities. But many of the former had access to education and other modern inputs. So they had some preparation to deal with the changes. Most traditional tribes, on the contrary, lived on mineral and forest rich land that the colonialist required as raw materials. That turned the imposition of the formal system on the informal societies into an unequal encounter. Land alienation from the traditional to "modern" communities was a consequences since the latter were unable to deal with the changes imposed on them.

This unequal encounter continues to be the basis of a disjunction and of conflicts between the two systems because the colonial laws continue to be in force in the country. One of its consequences is environmental degradation. The legal system that recognises only individual ownership is a major cause of land loss and environmental degradation. Since the CPRS are not recognised as their sustenance, the communities depending on them cannot prevent outsiders from encroaching on that land.

For example, in Tripura in North Eastern India, the tribal proportion has declined from 58 per cent in 1951 to 31 per cent in 2001 because immigrants have encroached on 60 per cent of their community owned land with the help of individual-based laws. Equally important is loss of forests which catered to many needs of the tribal and other rural poor communities. The state handed many of them over to industry as raw material. They were treated as sources of profit and destroyed with no concern for their dependants or for conservation. That impoverished people (Gadgil 1989.

The third source of land loss is acquisition for development projects. The law that empowers the state to acquire land recognises only individual ownership. More than 25 million hectares have been used in India for such projects 1947-2000, around 14 million of them forest and other CPRs. Their inhabitants, most of them tribal and other rural poor like fish and quarry workers are considered encroachers and are not compensated and often not even counted among the displaced (Fernandes 2008: 92).

Often records of the CPRs are not kept since they are considered state property and their inhabitants are encroachers. For example, according to official accounts, in Assam the state used 159,017.37 hectares of land for development projects and displaced 343,262 persons from them 1947–2000. The reality is 1.9 million persons displaced from 567,281.29 hectares (Fernandes and Bharali 2006: 107). More than 1.5 millions displaced persons and 410,261 hectares were not counted because according to the law these CPRs are state property and their inhabitants are encroachers with no right to live there.

The Vicious Circle

One can mention many other modes of land alienation. The above examples are given only to show the processes that lead to alienation of the people's livelihood. Because of the unequal nature of the encounter, also the reaction of these systems to the problems that the process causes differs.

That too is based on their worldview on land and the natural resources all of whose dependants feel the negative impact of the transition from the traditional to the modern economy and new values. But the rural poor, particularly the tribes and other forest dwellers feel its impact more than the remaining groups do because it is an attack on their tradition of judicious use of resources and on the systems they had developed to manage land, forests and CPRs as their renewable sustenance.

Loss of their sustenance begins the vicious circle of impoverishment that forces the dependants of these resources to overexploit them and cause further environmental degradation and more poverty. As the former Brazilian President Fernando Henrique Cardoso (1998) said, the first danger to the environment from people's impoverishment is loss of biodiversity and linked to it, loss of the values through which the communities depending on it had managed the resource as renewable. Studies show that loss of this value system or ideology is basic to the vicious circle that leads to further environmental degradation.

But reaction to this process differs according to the class one belongs to and one's ideology or culture. To the urban middle class land alienation and environmental degradation are loss of their recreational spaces while to the rural, particularly tribal, communities it results in loss of their livelihood and consequent impoverishment from which follows further land alienation and destruction of more natural resources. Conflicts are a natural consequence of this contradiction. The first step of this process is impoverishment of the economic status they are reduced to by the alienation of their sustenance. It begins with landlessness. Then comes joblessness. For example, studies of families displaced by development projects show that in Andhra Pradesh in South India, the proportion of the landless rose from 10.9 per cent before the project to 36.5 per cent after it and in Assam in the Northeast from 15.56 to 24.38 per cent. Even among those who retain land, the average area owned declined, for example in Assam from 1.2 to 0.6 hectares per family. In every state most small and marginal farmers became landless, and medium farmers joined the ranks of small and marginal farmers. They also witnessed a decline in the support mechanisms such as the number of irrigation ponds and wells, poultry, cattle, and draught animals that used to supplement their agricultural income declined.

Joblessness is the next step. The land and other resources that are alienated from them used to provide them work. They lose this resource with no alternative to take its place. Joblessness resulting from it takes two forms. The first is lower access to work and the second is downward occupational mobility. In Andhra Pradesh, for example, 83.72 per cent of the land losers used to work on their land or elsewhere before its loss. After land loss access to work declined to 41.61 per cent. In West Bengal it declined from 91.02 to 53.18 per cent and in Assam from 77.27 to 56.41 per cent. The second is downward occupational mobility. In most states, more than 50 per cent of the land losers who were cultivators before it became landless agricultural labourers or daily wage earners after land loss.

Also displacement can continue as a result of environmental degradation. For example, a new industry often forces people to move out of its neighbourood because after its construction environmental or other consequences such as fly ash and dust generated by the thermal, aluminium, nuclear, cement and other plants destroy the land around it and render it unusable. Its dependants cannot sustain themselves on it and are forced to move out. Also the noise and dust pollution and constant blasts in the coalmines often force people to leave their homes.

Absorbing a New Culture

The changes do not remain external but enter the community itself through the internalisation of the dominant culture. The major change is the culture the community in general and its elite in particular internalise, of viewing their sustenance as commodity alone. It is seen firstly in the demand the leaders make that individual ownership become the norm in their communities. For example, in the Garo tribe of East Garo in Meghalaya in Northeast India the leaders accepted the culture of individual ownership in the 1980s. A study two decades later shows that 30 per cent of the tribal families in this district had become landless since their elite had monopolised much of their land. These changes also have gender implications. As stated above, even the matrilineal societies are patriarchal. Their leaders absorbs the culture of greater patriarchy and express it in their land relations. That can be seen among the Garo who are a matrilineal tribe but individual ownership is through men. Among the Khasi of Megalaya who too are a matrilineal tribe, the male leaders who control the village council exploit their power to their own advantage and turn community-owned land into their private property (Mukhim forthcoming). Such change of gender attitudes is seen in other tribes too in the manner in which men interpret their customary law and property relations in their own favour.

Communities thus deprived of their resources absorb the same culture in another form. The first is the vicious circle of viewing their resources as a commodity alone. Once they are deprived of their resource and are impoverished, for sheer survival they overexploit the same resource for an income. For example, studies in all the tribal areas show that once they lose their land, the deprived families fall back on their forests that they had preserved for centuries and cut trees for sale as firewood or timber, and cause more deforestation. The second to view their own bodies as a commodity. For example, 49 per cent of the families displaced by development projects in West Bengal and 56 per cent in Assam pulled their children out of school in order to turn them into child labourers. Women began to view their bodies only as a source of income. Because of it prostitution grew enormously among the families that had lost their land. All these instances point to a major change in subaltern culture.

These communities lose hope in their future and think only of the present. As a result, children who are an asset for the future become commodities only for the present and are used as a source of income for survival. The same is the view of women's bodies. In other words, women and children become commodities more than men do.

This chapter which is an overview of the changes in land relations, has shown the new culture that grown as their result. It shows that imposition of another culture on a traditional group can result in a culture that is destructive of a community in general and of women in particular. The solution is not either going back to their tradition by opposing modernisation or absolutising the modern system. One cannot prevent all individual ownership either. One has to find an alternative in beginning with the tradition community values and combining them with the traditional community. Tradition has to be modernised and not replaced completely.

FOREST LAND MANAGEMENT

DEVELOPMENT OF FORESTRY

The Forestry Department is one of the earliest government agencies established in Brunei Darussalam. Forests played a significant role in the early economic development of the country, particularly during the pioneering exploration for and early production of oil, where considerable quantities of timber were required for construction purposes, necessitated the creation of a Forest Office in March 1933, based in Kuala Belait.

During the early days of its establishment, the primary role of the Forestry Department was focused on managing the forest resources for its economic benefits – managing the harvesting operations and collection of revenue from timber and other forest products, rather than managing the forest resources for its environmental benefits.

FOREST RESOURCES

Under the present management, realizing that the forest resources play a critical role in providing not only economic but also social and environmental benefits; as well as the needs to conserve and protect the forest resources and its biodiversity from the threats of deforestation and forest degradation, the Forestry Department shifted its forest management emphasis. The present forest management put emphasis not only on economic gains, but at the same time, putting equal emphasis between economic and the needs to protect and conserve the forest resources and its biodiversity, within the context of sustainable forest management. With Sustainable Forest Management, the Forestry Department is committed to support the national aspiration to diversify the country's economy, and at the same time, the Department is also committed to conserve and protect the forest resources and its biodiversity from being destroyed.

This is the challenge that the Forestry Department has to address in a wise and sound manner. In addition to this, the support from other government agencies and various stakeholders also play a critical role in determining the success of achieving this endeavor.

MANAGEMENT AND STRATEGIES IN FORESTRY DEPARTMENT

This chapter will highlight the management initiatives and strategies implemented by the Forestry Department in its effort to ensure the continued development and management of forest resources in a sustainable manner. Basically this chapter will briefly elaborate the policies and legislations that govern the management and administration of our forest resources as well as various initiatives, programmes and projects implemented to conserve and protect these resources.

This chapter will also highlight some of the challenges and issues faced by the Forestry Department in the management of these resources and how the Heart of Borneo initiative can be seen as an important tool in providing opportunities for a holistic approach in the management of the forest land and complementing the already existing efforts implemented by the Forestry Department in promoting sustainable forest management.

THE FOREST RESOURCES AND AREA

Brunei Darussalam is one of the countries in this region with the highest percentage of forest cover. It is estimated that about 76 per cent [438,000 ha] of the country's total land area is still covered with forest.

Forest Types and Its Percentage Distributions

The forest type of Brunei Darussalam is divided into two broad categories, namely the swamp and hill forests. The swamp forests are those which occur in low-lying lands which are subject to tidal, seasonal, or continuous flooding and inundation by water. This group composed of mangrove, freshwater swamp and peat swamp forests. The hill forests on the other hand are those that occur in generally high and dry grounds and this group composed of tropical heath forest (*kerangas*), mixed dipterocarp and montane forests. The descriptions of the forest types and its percentage distributions throughout the country.

MANAGEMENT OF THE FOREST RESOURCES AND FOREST LAND

The Vision of the Department is to achieve "Excellence in Tropical Forestry Management". This Vision encompasses the complete spectrum of forestry activities where Brunei Darussalam will strive for excellence, and by doing so, set an example of excellent and proper management of tropical forests to the rest of the world. The approach emphasizes on the needs to balance out development with environmental conservation and at the same time exploring and utilising the available forest resources for socio-economic benefits, within the context of sustainable forest management. The Vision as well as the Mission statement to support the Vision set out by the Department are as follows:-

- Vision, "Excellence in Tropical Forestry Management"
- Mission, "To develop, conserve and manage the forest for social, economic and environmental benefits, for the people through Sustainable Forest Management"

The vision and mission are set to accomplish the following strategic objectives, referred to as the "forestry excellence agenda", which focus on the importance of forest in satisfying the social, economic, as well as conservation needs, namely:-

Forests for Posterity and Prosperity

With the objective of perpetuating them as the country's natural heritage and the key to continued prosperity. This strategic objective is more focus on preserving and conserving the pristine forest for the benefit of the future generation.

Forests for Sustainable Production

This is aimed at making and keeping the forest resources productive in a sustainable, optimal, and environment-friendly way. This is to ensure continuous and sustainable supply of timber to cater for the domestic timber demand. At the same time, the development of forest plantation will continue to be pursued in order to reduce dependency hence the pressure on the natural forest for timber supply.

Forests for Economic Strength

This is intended to maximize the contribution of the forestry sector into the economic diversification efforts of the country. Main focus under this agenda is to further increase the contribution of the forestry sector to the country's Gross Domestic Product through the development of down-stream industry and value-added products. This agenda also calls for implementation of low impact utilisation of forests resources as well as the development as eco-tourism and biotechnology industry.

Forests for Public Involvement and Enjoyment

Designed to provide recreational opportunities and foster nature education among the people. Main focus under this agenda is to develop potential sites as forest recreation parks and National Park. The Department's role in this area is to build and upgrade park infrastructures and facilities as to make the parks more convenience and attractive to visitors.

Forests for International Prestige

For the purpose of making Brunei Darussalam a world class model in the field of forestry. With the ability of Brunei Darussalam to keep a large portion of its forest still intact and in its pristine state, is something that could possibly promote Brunei to the world as one of the country that is committed in conserving its green heritage and has sound forestry management practices.

THE MANAGEMENT OF THE FOREST RESOURCES AND FOREST LAND

Laws of Brunei mandated the Forestry Department to be the primary government agency responsible in the management and administration of the forest resources of the country.

The Act provided the basic law for administration of the forests, reservation of forest lands, harvesting of forest produce, grant of customary gratuitous rights to forest-dependent inhabitants, stipulating penalties for violations, and prescribing forest royalty. In addition to this legal instrument, the management and administration of the forest resources is also guided by the 1989 National Forest Policy and the long-term forestry sector strategic plan [20 years plan for the period of 2004-2023]. Relevant legislations that strengthen the management of forest lands are the Land Code and Land Acquisition Act; Wildlife Act; Town and Country Planning Act; Antiquities and Treasure Trove Act; and Wild Flora and Fauna Order 2007.

The provisions in the Forest Act provides the Forestry Department with the power to manage and administer the forest resources of the country, however, as far as the overall planning, development and management of the land with existing forests remains under the jurisdiction of the relevant government agencies and individuals, where the land are gazetted or allocated to.

The Forestry Department has the sole right to the overall planning, development and management of forest land which has been gazetted as Forest Reserve. However, while the overall responsibility of managing the forested land on the State Land, Reserved Land and Alienated Land falls under the jurisdiction of the respective government agencies or individuals. The role of the Forestry Department in these areas is limited only in administering the harvesting and utilisation of timber and other forest products therein as provided for under existing forest laws, rules and regulations.

Forest Laws and Relevant Legislations

Apart from the Forest Laws and relevant legislations, there are also other regulations and guidelines that regulate the forestry harvesting operation. Brunei Darussalam has its own harvesting guideline in order to achieve the following objectives; to maintain and increase the timber production capacity in the designated production forest areas; and to balance the timber harvesting and biodiversity conservation *i.e.* extracting commercial timbers while meeting the need for forest conservation. The guideline covers every aspect that is considered crucial in maintaining the sustainability of the forest, *i.e.* from the construction of man forest roads to logging procedures and post-harvest reporting.

The guideline is primarily aimed in facilitating the day-to-day supervision and logging activities monitored by field officers, and ensuring that all such activities are in accordance with the Brunei Selection Felling System (BSFS). BSFS involves pre and post-logging assessment of the timber stand. Trees to be cut and harvested as well as trees to be left behind to constitute the next timber crop are marked. Felling includes commercial species and sized which are governed by a set of diameter limits.

To safeguard these forest resources from any illegal activities, the Department conducts regular forest patrol via air, land and water. This forest patrol activities are being carried out either as part of the departmental enforcement unit routine forest patrol activities or as part of the joint forest patrol operations with the relevant government enforcement agencies such as through joint operations under the *"Jawatankuasa Protap Salimbada"* - members of which comprises of Royal Brunei Police Force, Royal Brunei Armed Forces, Survey Department, Land Department and Forestry Departments.

The Management of the Forest Reserve

Out of 438,000 ha of the forested area, only about 235,520 ha [40 per cent of the country's total land area] has been gazetted as Forest Reserve. In line with the national forest policy, the Department aims to gazette an additional 15 per cent of the country's total land area as Forest Reserve, making the total Forest Reserve in the country to 55 per cent of the country's total land area. For management purposes, this Forest Reserve is classified according to its functional categories namely protection, production, conservation, recreation and national park. The functional description, detail breakdown of this classification and the location of gazetted and proposed Forest Reserve throughout the country.

Forest Plantation Area in the Inter Riverine Zone

The Forestry Department aims to develop at least 30,000 hectares of timber plantation to support the demand for wood raw materials of the country's forest-based industry on a sustainable basis. This initiative also intends to ease the pressure of natural forest utilisation and confine timber production in plantation forests. The large-scale timber plantation establishment was based on several trial plantings of different timber species conducted since the late 1960's. A comprehensive study conducted in 1984 recommended the development of the sawtimber plantation development within the Inter-Riverine Zone (IRZ) located between Tutong and Belait River

The private sector plays an important role in forest plantation development. They are being tapped by the Government as partners in this endeavor, being a project implementor on the ground. During the process, the private counterpart shares the responsibility of managing the resource and at the same time learns and develops their technical skills in timber plantation establishment under the guidance of the Forestry Department.

LAND USED FOR RECREATION AND WILDLIFE AREAS

Special uses include rural transportation; rural parks and wildlife; defence and industrial uses; and farmstead, farm roads/lanes, and other onfarm uses. These special uses increased from 85 million acres (4 per cent of the land area of the contiguous 48 States) in 1945 to 153 million acres (8 per cent) in 2002. Land in transportation uses (highways and roads, railroads, and airports in rural areas) increased by 4 million acres (17 per cent) between 1945 and 1982. Transportation uses declined by about 0.5 million acres from 1982 to 1992 due to the abandonment of railroad facilities and rural roads, and the classification of some transportation uses as urban areas.

Land used for recreation and wildlife areas (Federal and State parks, wilderness areas, and wildlife refuges) expanded 344 per cent from 1945 to 2002 (an increase of 78 million acres). The increase came mostly from conversion of Federal lands, previously in forest and grassland pasture and range. Land in defence and industrial uses declined by 10 million acres (40 per cent) from 1945 to 2002.

Farmsteads, farm roads, and other farm uses declined by 4 million acres (29 per cent) between 1945 and 1997. This decline reflects trends towards fewer farms and larger, more consolidated farms, as well as an increasing tendency for farm households to live off the farm. In response to expanding U.S. population, land in urban uses—including homes, schools, office buildings, shopping sites, and other commercial/industrial uses—increased from 15 million acres in 1945 to 25 million acres in 1960, 47 million acres in 1980, and 59 million acres in 2002. While the U.S. population nearly doubled, the amount of land urbanized quadrupled. However, urban uses still comprise only 3 per cent of the total land area of the contiguous States.

Miscellaneous other land uses decreased from 1945 to 1964, and have since trended upward, showing a 54-per cent increase from 1964 to 2002 reflecting improved data and reclassification of grazing and forest lands. These uses include marshes and open swamps not included in other major land uses, bare rock areas, deserts, some rural residential areas, and other uses not inventoried. Wetlands are defined by soil and hydrological characteristics and may occur on land in many different uses.

REGIONAL CHANGES IN LAND USE

While land in every use occurs in all 10 regions of the contiguous States, some uses are more concentrated in some regions than in others. Regions with the largest cropland acreage are the Northern Plains, Corn Belt, and Southern Plains. Grassland pasture and range is concentrated in the Mountain and Southern Plains regions. Acreage in forest use, special and miscellaneous other uses is highest in the Mountain region.

The Northeast, Appalachian, Southeast, Delta States, and Lake States regions lost cropland between 1945 and 2002. The largest increases occurred in the Northern Plains and Mountain regions, with smaller increases in the Corn Belt, Southern Plains, and Pacific regions. Western increases may have resulted in part from federally subsidized irrigation water.

Nine of the 10 regions lost grassland pasture and range between 1945 and 2002. While grassland pasture and range increased 11 million acres (10 per cent) in the Southern Plains, the Northeast region lost about 70 per cent of its grassland pasture and range, and the Appalachian and Lake States regions lost more than 50 per cent. The Northeast and Appalachian regions saw the reforestation of grassland, loss of some grassland to urbanisation, and concentration of the dairy industry. Decreases in the Corn Belt, Northern Plains, and Mountain regions were likely associated with the conversion of some grassland pasture and range to cropland.

Cropland Use and Federal Programmes

While total cropland acreage has varied up and down and generally declined since 1969, greater shifts have occurred between cropland used for crops and cropland idled, mostly because of Federal programmes. Cropland used for pasture has exhibited less variation in acreage than cropland idled. Most cropland used for crops is harvested, but typically 2-3 per cent experiences crop failure and 5-10 per cent is cultivated summer fallow. In 2002, farmers harvested one or more crops on an estimated 307 million acres of cropland, down 4 per cent from 1997. About 8 million acres of the total harvested were double-cropped. When double-cropped land is counted twice, total acres harvested rise to 315 million acres.

Table. Major uses of cropland, 48 contiguous States, 1992-2004

Cropland Million acres	1992	1997	2002	2004[1]
Cropland used for crops[2]	337	349	340	336
Cropland harvested[3]	305	321	307	312
Crop failure	8	7	17	9
Cultivated summer fallow	24	21	16	15
Cropland idled by all Federal programs[2]	55	33	34	35
Annual programmes	19	0	0	0
Conservation Reserve Programme	35	33	34	35
Total, specified uses[2,4]	392	382	374	372

Note:

[1]Preliminary, subject to revision.

[2]Breakdown may not add to totals due to rounding.

[3]A double-cropped acre is counted as 1 acre.

[4]Does not include cropland pasture or idle land not in Federal programmes that is normally included in the total cropland base.

Cropland used for crops was at a record high of 387 million acres in 1949, when no acres were idled by Federal programmes. In 1972, cropland used for crops was near a record low of 334 million acres as Federal programmes idled 61 million acres. Cropland used for crops climbed to 387 million acres in 1981, when Federal programmes idled no cropland, and dropped to 333 million in 1983, when Federal programme set-asides reached a historic peak of 78 million acres under the Payment-in-Kind (PIK) programme. The Federal Agricultural Improvement Act (FAIR) of 1996 eliminated all Federal acreage reduction programmes other than the Conservation Reserve Programme (CRP)

Between 1983 and 2002, cropland used for crops increased overall, while total acreage idled by Federal programmes decreased to 34 million acres. From

1997 to 2002, cropland used for crops declined from 349 to 340 million acres, while acreage idled under CRP increased by about 1 million acres. Between 2002 and 2004, acreage in CRP increased by an additional 1 million acres, while cropland used for crops declined by about 4 million acres.

The 14-million-acre drop in harvested cropland between 1997 and 2002 was coincident with a decrease in cultivated summer fallow and an increase in failed acres due to widespread drought. Crop failure occurred on 17 million acres, over 5 per cent of the acreage planted, in 2002. This failed acreage was the largest since 1956. The use of summer fallow has been decreasing since the late 1960s, and stood at 15 million acres in 2002, down from 42 million acres in 1969. Four crops—corn for grain, soybeans, wheat, and hay—accounted for 80 per cent of all crop acres harvested in 2002. The additional 17 "principal" crops accounted for another 15 per cent of harvested area. Vegetables, fruits, nuts, melons, and all other crops accounted for 4.5 per cent of crop area harvested in 2002.

URBANISATION OF AGRICULTURAL AND OTHER RURAL LAND

Cropland conversion to urban uses is largely irreversible, so it is important to know the rate of conversion and how much of the loss is replaced from other land uses. Excessive loss of cropland to urban uses could lessen the production of food and fibre and the supply of rural amenities, such as open space, watershed protection, and rural lifestyles. A variety of Federal, State, local, and private programmes address such concerns. Although urban land constituted less than 3 per cent of the U.S. land area in 2000, 79 per cent of the population lived there.

Even large percentage increases in urban area would amount to small decreases in rural area since it is so vast. The rate of expansion (by decade) of urban area has declined from 39 per cent during the 1950s, to about 36 per cent during the 1960s and the 1970s, and to 18 per cent in the 1980s. According to the Census Bureau, urban area was 59 million acres in 2000, just 7 per cent above the previous estimate for 1990 (DOC/BOC, 2002). However, the Census Bureau adopted a new definition of urban area for the 2000 Census, improving the precision of urban area measurement but making it more difficult to compare urban area before and after this year. If urban area for 1990 is recalculated using the 2000 definition, the Census Bureau's estimate falls to 51 million acres, implying a 15-per cent increase in urban area from 1990 to 2000.

The National Resources Inventory (NRI), conducted by USDA's Natural Resources Conservation Service (NRCS) in cooperation with Iowa State University, is an alternative source for estimates of urban and rural areas. The NRI uses a consistent definition for built-up areas, though it differs from the definition used by the Census Bureau. According to the NRI, "developed land," which includes large and small urban and built-up areas as well as rural transportation land, totalled 107 million acres in 2002 in the contiguous United States. The NRI indicates that developed land increased by 14 million acres (19 per cent) over 1982-92 and 21 million acres (24 per cent) over 1992-2002.

Land converted to urban uses comes from several different major land uses. The NRI indicates that 20 per cent of new developed land came from cropland between 1997 and 2002. Prime cropland— land that has the best combination of physical/chemical characteristics for agricultural production—is converted to developed uses at about the same rate (5 per cent per year) as non-prime cropland.

About 21 per cent of rural non-Federal land is prime and 26 per cent of crop land converted to urban uses over 1997-2001 was prime. Rural land, defined as all land that is not urban, contains rural residential land, consisting of houses and associated lots. Non-farm rural residential area was estimated to be about 94 million acres in 2002, up from 56 million acres in 1980. The average rate of increase in rural residential land was 1.7 million acres per year from 1980 to 2002. Combining both rural and urban residential land, the total increase in residential area was about 2 million acres per year during this period.

FARM FORESTRY—A WIN-WIN SITUATION

Farm forestry or social forestry is a silent movement sweeping across parts of rural India which has the potential of creating another green revolution. Simply put, farm forestry is aboutcorporates promoting cultivation of trees as a crop by small farmers. Farm forestry was initially promoted as a NABARD assisted project during 1987-95. Now the paper industry is using this tool effectively to get its raw material without harming forests and at the same time generating income and employment for the farmers. This helps the environment, thecorporates and the poor at the same time, which is a rare proposition.

The Need

As nearly 50 to 60 per cent of the paper mills in the world are wood based, forests have a very important role to play in this industry. Forests account for nearly 30 per cent of the world's land surface. The shrinking forest area (due to the increasing population and use of wood for various applications) is a cause of concern for everyone.

The paper industry accounts for nearly 12 per cent of the world's wood harvest (including the harvest for fuel purpose). If the wood harvest for industrial purpose alone is considered, the paper industry accounts for nearly one fourth of the total industrial consumption.

Forest cover in India is much lower than the world average, and an increasing rate of population promises that the pressure on forests will increase manifold and reach unsustainable levels.

The Solution

To meet the increasing need for raw material without harming the environment, some large paper companies have turned to farm forestry. Companies are working with NGOs to identify poor households in rural and tribal areas owning wastelands and organising them into self-supporting groups. These groups are provided with good planting stock of pulpwood species and other local species that meet domestic, fodder, fuel and nutrition requirements. They are also provided with technical and financial assistance and an assured buy-back of wood. Technical assistance includes services like land development, planting of saplings and plantation maintenance. Farmers plant trees on their land as a crop or on the periphery of their farmland as an additional source of income. A project in Andhra Pradesh has even gone a step further where the government is ensuring access to land for the poor. Some of this land is completely scrub, lying unutilized for generations. This project has also brought about pooling of the small land holdings of the poor to form land banks.

Bio-technology comes in handy

Companies are using biotechnology to leverage the development of high yielding clones of pulpwood species as well as to create desirable characteristics in such trees. For instance, one clone developed by ITC has a root system that does not go beyond 1.5 meters in depth, thus minimizing access to precious ground water. Some clones have productivity rates six to nine times those of conventional seedlings and some are site-specific clones adapted to problematic alkaline and saline soils.

Micro benefits for the backward

Farm forestry has a substantial employment generation potential and this is a huge positive for a populous country like India with high unemployment rates. A significant part of the land under such projects is wasteland owned by marginal farmers, many in tribal areas. This project increases incomes of poor households in some of the most backward districts of the country. ITC alone has created around 3 lakh jobs through its farm forestry project, which had covered around 29,000 hectares of land by 2004. Ballarpur Industries Limited (BILT) and JK Paper had also covered 21,900 hectares and 27,000 hectares of land respectively under their farm forestry programmes till 2004.

There are macro benefits as well

Apart from increasing the forest cover, farm forestry reduces pressure on public forests by providing fuel wood to poor households from their plantations. It also contributes to in-situ moisture conservation, reduction in air pollution, groundwater recharge and reduction in top-soil losses. Depleted soil is also enriched due to leaf-litter from plantations. This might lead to a decline in fertilizer and pesticide consumption in the future, thus reducing thc pollution of groundwater sources. Over the years, this alternative source of employment and income for a section of the marginal landless farmers and ofVadgam, a small village in Gujarat, has resulted in a positive change for them, through a partial disintegration of negative social patterns of traditional land ownership. Since it is difficult to find employment, the villagers rely on their land, however small and unproductive it may be. They find work as agricultural labourers only during the peak season and remain without work for several months of the year. Thus, they require both secure employment and regular income to keep out of the clutches of money-lenders.

Economic benefits

To top it all, farm forestry programmes have huge carbon sequestering potential, thus creating additional economic benefits in the form of carbon credits (*refer I-Mag issue Sep 2005 "Demystifying Carbon Credits"*). According to an estimate, ITC's plantations alone have the potential to sequester 18 lakhtonnes of carbon, reducing 67 lakh tonnes of carbon dioxide, thus creating a carbon credit value of ₹ 25 crores per annum approximately. However, it needs to be ensured that the economic benefits of these carbon credits accrue to the farmers who own the land, rather than to the corporates driving the initiative. This will be a bonus for small and marginal farmers of the country, 40 per cent of them living below the poverty line.

Challenges

The corporates would have to disseminate information about the benefits of land pooling for marginal and scattered land holdings, which might not be readily accepted by the farmers. The cost benefit analysis for the corporates themselves might also not favour such an undertaking unless the benefits of carbon credits, conformance to environment regulations are factored in.

FARM FORESTRY: LAND AVAILABILITY, TAKE-UP RATES AND ECONOMICS

The target set by the Irish Government is to increase the forest cover to 17 per cent of the land area of the country by the year 2030. This involved an annual afforestation level of 25,000 ha to the year 2000 and 20,000 ha thereafter to the year 2030. The target was almost achieved in 1995 but planting levels have fallen far short since, even though substantial financial incentives were put in place.

IMPLICATIONS

Failure to reach the overall target by 2030 will result in failure to reach the critical mass of forest output required to support a viable timber industry.

Since the employment creation potential of large timber mills is not very high, it is essential that significant value added accompanies whatever timber output is achieved in order to boost employment. In the absence of achieving the afforestation targets listed in the Strategic Plan for Forestry, the contribution of the forestry sector to the growth targets of the National Development Plan will be less than expected and the objectives set out regarding forestry in the CAP Rural Development Plan will not be achieved.

Background

The study objectives were to examine the factors:

- Influencing land availability for forestry,
- Slowing down or accelerating the rate of afforestation and
- Affecting the economic return from farm forestry.

The amount of land becoming available for afforestation via the market has declined substantially in recent years. Land transfers are via gift or inheritance in the vast majority of cases. Land becoming available (non market) for forestry at time of change of ownership was not examined but a survey of landowners opinions was used to study the factors influencing decisions to plant. In this study the economic returns from forestry were compared with the returns from existing farm enterprises using linear programming techniques. The technique as applied in this study is aimed at maximising total farm gross margin from all available resources thereby arriving at the optimum mix of enterprises (including forestry). The main assumptions of the economic study were as follows:

- Returns to forestry were calculated using a 10 per cent discount rate.
- The gross margin system is used to compare returns from forestry with agricultural enterprises.
- No economic value was given to the amount of carbon dioxide fixed by growing trees, and
- Farm labour was valued both at minimum wage and at average industrial wage rates.

The discount rate normally used by those in the forestry sector is 5 per cent. Fixed costs are proportionally much higher in agriculture than in forestry. Forestry is long-term and all costs are variable in the long term so perhaps a net figure might be more suitable for comparison purposes.

Findings

Land Availability and Take-up Rate

Landholders attitude to forestry are becoming more positive. The unfavourable responses to forestry were based mainly on pragmatic rather than cultural issues, *e.g.* small farm size, land quality, competing EU payments. The main reason for planting was the very favourable income from forestry premium payments on land that had limited other uses. Forestry on good farmland was not favoured but was a definite competitor to agricultural enterprises on land marginal for farming and/ or situated some distance away from the main farmyard. Afforestation was mostly associated with large farm size. The study did not examine the question of cutting back in farm enterprises (livestock numbers) which is often associated with old age and/ or ill health. The demographic profile of those in the sample with farm forests was the same as in the National Farm situation.

Economic Returns

Opportunities for off-farm employment and levels of remuneration can determine the choice of farm enterprise and the rate of afforestation. Off-farm earnings at or near industrial wage rate levels may result in increases in the area planted often to the exclusion of cattle and sheep enterprises at individual farm level. Since the decision to plant is for all time, economic factors alone are often not the only criteria for decision-making. Where off-farm jobs are not available, forestry will compete with farm enterprises only on farms where there is land surplus to that required to maximise REPS, compensatory allowance, and extensification payments.

However, even where no off-farm jobs are available, forestry will always compete with drystock enterprises operated at moderate and low levels of efficiency. The study gives a clear message that a switch to forestry can improve overall farm and household income for many farmers depending on their circumstances. The findings of the study are important pointers for Teagasc staff involved in the Opportunities for Farm Families Programme.

FARM FORESTRY: RESPONSE TOWARDS THE SCARCITY

The real crisis faced by the Third World Countries where firewood, crop residues, kerosene being the popular form of energy for cooking, Eckholm wrote "the people are scrambling to cook dinner under deforestation". However, in the wake of deforestation, villagers coped with increased scarcity in various ways. Introduction of tress in the private land is widely pursued by the farmers elsewhere. A study, using aerial photographs between 1964 and 1988, observed a tremendous increase of private trees in the hills of Nepal. It is postulated that this increase is mainly due to the decreasing access and unavailability of forest resources.

This shift of trees has occurred elsewhere. Pakistan gives example of Hazara forests managed by villagers in frontier tribal areas under weak state control. An African study shows " rural households are planting trees in the wake of labour shortages due to off-farm income opportunities.

Farmers favour multipurpose trees over trees that supply only firewood, as cheap substitutes for firewood are available. Similarly planting trees only for fodder alone is not common, but existing fodder trees are retained. Thus introduction of trees in private land is a strategy adopted by villagers to cope with emerging biomass scarcity. Timber trees are stores of values for household savings that do not put demand on household's scarce cash resources. However, tree growing is determined by the farmer's livelihood strategies and resource base. Farmer's decision regarding tree growing depends on:

- Declining access to the wood land,
- Loss of communal lands or restrictions on forest access
- Increasing demand for forest products
- Increasing demand from the market.

A survey from western Kenya shows that the "poor" prefer firewood. The "average" had a higher portion of fruit and timber while the wealthy invests heavily in fencing. A study on farmer's willingness to grow trees in Gunung Kidul opts for a different reason. As the farmer's willingness to grow trees depends on many factors. An increase in the productivity of staple crop is an important factor permitting farmers to plant trees. Government policy can also create a favourable market trend for growing trees

CONSTRAINS OF TREE GROWING

The following are the main constraints for tree growing by the rural poor:

1. Economic: trees can't compete with other uses for land, labour and capital.
2. Social/cultural: shelter to evil spirits, harbours insects etc.
3. Site: condition not meeting requirement for trees
4. Land use rights tenure security
5. Red tape(ism): Permits for utilisation

The above constraints affect the farmer's willingness to grow trees while Gregersen *et al.*, 1989 gives some reasons with regard to flexibility that can be (dis)incentive for farmers to grow trees:

- Multiple tree outputs (adjust as the market change)
- Flexibility in harvesting time
- Low liquidity in early years reduces flexibility.

Besides the above, the higher rate of discount in case of developing countries discouraged farmers from introducing trees in their farmlands.

Social forestry interventions in south Asia intended to involve local communities, whether it is agro forestry, farm forestry or community forestry needs well planned transformation of a dynamic inter-relationship among community, natural environment and the state. In certain situations, the forest-community dependencies are deep rooted and need careful consideration while designing such development interventions. The theoretical basis for planning such transformations is more complex than project planners and bureaucrats often consider. Most of these forestry interventions went through a number of paradigm shifts, beginning with a top down conventional model approach, to the most recent approach wherein emphasis is laid on local participation. There is a lack of proper understanding by policy makers and foresters, and inadequate theoretical basis in these approaches. Analysis of the approaches may provide explanation for the poor performance of most social forestry projects in south Asia. Different views exist about the poor outcomes and the basic reasons for such outcomes.

Earlier studies that addressed issues of social forestry in south Asia tended to neglect local conceptualisations of natural resources, because their methodologies excluded local histories, local socio-political process and issues of institutional change. Studies in the recent years have started paying attention towards the analysis of local livelihoods, the local complexity of natural resource management and the long traditions of local resource management and conservation (Blaikie and Brookfield 1999; IIED 1998; IDS 1999).

This study will attempt to analyse some of the theoretical contents of the social forestry projects in south Asia and related policies in order to make them explicit and the factors responsible for the poor performance of the social forestry projects with regard to equity, environmental rehabilitation and sustainability.

The present study draws on the authors' research and field experience in social forestry in South Asia (India, Nepal and Sri Lanka) since the mid 90s. Some data was also gathered during specific reviews of three social forestry projects in India, Nepal and Sri Lanka funded by different international donor agencies.

10

Forest Policy

INTRODUCTION

Before 1865, forest dwellers were completely free to exploit the forest wealth. Then, on 3 August 1865, the British rulers, on the basis of the report of the then-superintendent of forests in Burma, issued a memorandum providing guidelines restricting the rights of forest dwellers to conserve the forests. This was further modified in 1894.

The sole object with which State forests are administered is the public benefit. In some cases the public to be benefited is the whole body of tax payers; in others the people on the track within which the forest is situated; but in almost all cases the constitution and preservation of a forest involve, in greater or lesser degree, the regulation of rights and restriction of privileges of users in the forest areas which may have previously been enjoyed by the inhabitants of its immediate neighbourhood.

This regulation and restrictions are justified only when the advantage to be gained by the public is great and the cardinal principle to be observed is that the rights and privileges of individuals must be limited otherwise than for their own benefit, only in such degree as is absolutely necessary to secure that advantage. In actual practice, however, all these pious declarations were set aside whenever they came in the way of British interests. For example, forests in Nagaland and the Terai were unscrupulously cut to meet the increasing demand of wood during both world wars.

The National Forest Policy of the Government of India (1952) is an extension of this policy. This policy prescribed that the claims of communities near forests should not override the national interests, that in no event can the forest dwellers use the forest wealth at the cost of wider national interests, and that relinquishment of forest land for agriculture should be permitted only in very exceptional and essential cases. The old policy of relinquishing even valuable forests for permanent cultivation was discontinued and steps to use forest land for agricultural purposes were to be taken only after very serious consideration. To ensure the balanced use of land, a detailed land capability survey was suggested. Conservation of wildlife was to be regularized. The tribals were to be weaned away from shifting cultivation. The concept of "national interest" has been applied in a narrow sense. A welfare state cannot have a basic contradiction between local and national interests. As the analysis that follows will show, in the implementation of the forest policy, "national interests" remained confined to augmenting revenue earnings from the forests. Whenever the interests of the local people or ecological considerations hampered possible revenue from forests, the forest department pushed them aside on the pretext of broader national interest.

Forest dwellers have been dissociated from the management and exploitation of forest wealth. The British contractual system that still exists in many states has resulted in unscrupulous exploitation of the local people and of the natural vegetation and wildlife that the forest policy was intended to conserve. Development programmes-construction of roads and availability of educational, medical and housing facilities-have allowed economically viable outsiders to enter forest regions. In order to make quick profits, they have exploited the forest dwellers, displacing them from their land and making them bonded labourers. Except for the states of Haryana, Punjab, Rajasthan, Tripura, Andaman and Nicobar Islands, Goa, Daman and Diu, the government has been earning huge net forest revenues. Over the years, this revenue has been increasing.

On the other hand, except in a few places, the condition of the forest dwellers has been deteriorating. The government's various development programmes and the tribal welfare schemes have by and large failed to make any dent in the deteriorating condition of the forests and forest dwellers. The crux of the problem lies in misdirected policy and its half-hearted implementation.

RESTRICTIONS AND BACKBONE ON FOREST DWELLERS

The backbone of forest dwellers' economy is the vegetation found in the nearby forests. But the forest dwellers' rights to collect fuel, fodder and minor forest produce are very restricted. (In Arunachal Pradesh, the tribals have special rights to collect all forest produce and hunt and fish freely in all forests, whether reserved or unclassified. This concession is not found anywhere else). In almost all states, forest dwellers can collect forest produce, either free or at concessional rates, only from protected and unclassified forests. Even these are not legal rights, but are subject to various government restrictions and regulations. These facilities can even be terminated if the capacity of forests does not permit the exercise of the right. The right holders are also obligated to help the forest department prevent and fight forest fires and to help the staff in connection with forest offences. In West Bengal, since Independence, forests have been the bone of contention between the forest department and the forest dwellers, most of whom are tribals. In forest dwellers contend that they, by virtue of being native people, have the right to use forest trees for their livelihood.

One of their complaints is the cattle trenches dug around the forest areas prevent the free flow of water into their arable land. They were charged high rates for permits to collect minor forest produce, especially tussar cocoons. Grazing has been prohibited and their agricultural land has been recorded as forest land in the recorder of rights; later the tribals were asked to vacate the land. The forest department, on the other hand, blames the tribals for wanton and indiscriminate destruction of wilds vegetation and wildlife, sand urges the restriction of their rights over the forests.

The tribal welfare department, while conceding the need to preserve the forest, says that the tribals do have special claims over the forests and forest produce. A government study team found the attitude of the forest department in many states to be one of callousness, indifference and neglect towards tribals. For example, the forest department in Karnataka (then Mysore), in its enthusiasm to set up a game sanctuary in the state, drove out the local inhabitants without making alternate arrangements for their resettlement. The intervention of the social welfare department was of no avail. The roads constructed in reserved forests by the social welfare department of the state for the benefit of the tribals, on the express understanding that they would be maintained by the forest department, have been badly neglected. The forest department must be persuaded with great difficulty to release water from tanks lying in reserved forests for irrigating the tribal areas lower down. In Andhra Pradesh, too, extensive areas under cultivation were being converted into reserved forests. The forest department has extended its boundaries almost to the doorstep of many tribal villages, instead of leaving open a strip, at least one kilometre in width, between the villages and the forest boundary. Areas were also included in the forests which had no tree growth at all and heavy fines were being imposed on tribal encroachers.

In Tamil Nadu (then Madras), the study team found discrimination against tribals but not non-tribals in Wenlock Down. The latter were given land on pattas (land titles) either permanently or for long periods, but the Todas-local inhabitants-received only annual cultivation permits. The study team also noted that extensive areas now classified as forests were under unauthorized cultivation. Without rights of ownership, the tribals could not undertake soil conservation measures. Moreover, the forest department's practice of serving eviction notices to tribal encroachers and fining them operated as a great hardship to the tribals. Finally, the study team observed that tree pattas, which entitle the holder to the use of the fruits of trees, were issued in Coimbatore district to non-tribals living far away from the tribal settlements, in respect to mango and tamarind trees in the tribal villages.

O.P. Arya, in his interview with the officials of the forest department, voluntary agencies and forest dwellers, has tried to show the "bitter relationship between the forest administration and the local populace." The divisional forest officer he interviewed was "critical about the attitude of tribals towards the forest." His view was confirmed by the collector and the charge officer. On the other hand, the people-about 80 per cent-complained during the field survey about harassment by the forest officials. The privileged ones, however, had good rapports with the forest department.

The Irish missionary stationed in the area who, according to the author, gave "a very impartial view of the whole problem," said that "the forester used to take a lot of benefits from the tribals of the village. Some of them used to be employed in the forest and were paid much less than government rates. He used to take gifts in kind, too." Since the villagers depend solely on forests for their livelihood, the forester behaves like a ring-master and makes them do what he desires.

Forest Dwellers and Department

The various kinds of restrictions imposed on the forest dwellers virtually put them at the mercy of the forest department, especially lower-level functionaries. Illiteracy and poor economic conditions make them even more vulnerable. In some areas in Andhra Pradesh, the forest guards have their cut of minor forest produce earnings, and in some areas the tribals are often made to work without pay. In spire of the rest houses spread all over the block, the forest rangers or high level functionaries could not find it convenient to inspect the area.

This is not an isolated instance. The IAS probationers, during their survey of socioeconomic conditions of tribals in different states, noted different modes of exploitation by the forest guards and other lower-level functionaries who were invariably in league with the patwari and the local police constable. Forest dwellers have also become victims of the commercial exploitation of the forest. Since British rule, contractors have employed the forest dwellers to do unskilled jobs for low wages and in appalling conditions. In the late 1940s, Symington recommended the association of local inhabitants in the exploitation of forest produce. In 1946-1947, the Forest Labour Societies were initiated by B.G. Kher in the then Bombay state. The purpose of the Forest Labour Societies was "not only to give the Adivasi labourers full wages along with a share in the profit, but also to train them gradually to take up the responsibility of conducting forest and other business by their cooperative efforts."

The Bombay state experiment was complemented in the First Five Year Plan. The Second Five Year Plan recommended the establishment of Forest Labour Societies in an increasing number in all states. It recognised that "the manner in which the forest resources are exploited has a great deal of bearing on their welfare. In many ways the penetration of forest contractor into tribal economy has been harmful". The Third Five Year Plan stated that "development of forestry and forest industries is also essential for raising the income of the tribal people who live in the forest areas".

The Working Group of Welfare of Backward Classes for the Fourth Plan stated that the manner in which the existing forest policy is understood and implemented had placed the tribals completely at a disadvantage. The Resolution of the Central Board of Forestry reiterated replacing the contractors with Forest Labour Cooperatives. Gradually, the Forest Labour Cooperatives were set up in most of the states. But the contractual system of exploitation has not yet been rooted out. The Union Minister for Agriculture, Rao Birendra Singh, promised in the Parliament to end the contractual system of exploitation of forest wealth within three years. In fact, the forest dwellers' share as wage earners in cutting, logging and loading was significantly less than the ruling market price. The forest dwellers can gain from the forest wealth only when they are involved in its processing through cooperatives.

The forest department also employs labourers for a variety of tasks such as road repair, road making and silviculture operations. Workers are also employed for departmental construction-building bridges, culverts and causeways. This gives much needed additional income to the tribals. With the expansion and development of forest-based industries, the demand for the exploitation of minor forest produce has increased. Unlike the major forest produce, minor forest produce in several states is collected by the government agencies and in others by businesspeople. Both the forest department and businesspeople are exploiters. In Madhya Pradesh, trade in tendu, patta, timber, gum, harra, bamboo, sal seeds and khair is nationalized. The payment of wages in the collection of tendu leaves is on the basis of number of bundles collected. The rate per bundle is different in the bordering states of Bihar and Uttar Pradesh. For example, when in Uttar Pradesh the rate was higher by 20 paise (US 20) per bundle, a person could not collect more than two bundles a day. Just to earn 40 paise (US 4c) more, the tendu leaf collectors in Madhya Pradesh used to walk 12-20 miles during the night and sell the bundles in Uttar Pradesh anon.

In Orissa, the forest department has taken over the collection of tendu leaves. The forest region is classified into subdivisions and ranges. Each range has about 10 collection centers from which pluckers receive 10 paise (US 1c) for leaves. In this way the government earned annually ₹ 44 million (US $3.4 million) or royalty, and the Forest Corporation a profit of ₹ 20 million (US $1.5 million). But this profit had been made at the cost of the tribals.

Where the minor forest produce is not nationalized, businesspeople exploit the forest dwellers. In the Uttarakhand region in Uttar Pradesh, contractors employed local people at nominal wages, but made huge profits themselves. In

Chamoli district the Dashauli Gram Swarajya Sangh intimidated the local people about the prevailing market prices of the different herbs and mushrooms they had collected, and advised them to sell their herbs rather than collect them at nominal wage for the contractor.

Some tribal regions have no regular market. At the weekly mandis (marketplaces), private wholesale dealers buy the minor forest produce at very low rates and sell them at huge profits in the cities. Many times they pay the tribals in advance and later buy their goods at nominal costs. Since most of the tribals do not know how to count, the price that is promised often does not correspond to the amount they actually receive. Generally, the trader is a moneylender, too, who buys the produce from the tribals as repayment of a debt, but at incredibly low prices. The Tribal Development Corporations, Forest Corporations and some other cooperative agencies have taken up the sale of some produce, but the moneylender-cum-trader has yet to be eliminated by them.

LAND POLICY IN URBAN PLANNING

Since urban planning in India is largely concerned with development of land, it would be relevant to briefly consider how perceptions about land and real estate property have evolved. The Indian Constitution initially recognised 'to acquire, hold and dispose of property' as a fundamental right. Consequently when land was to be compulsorily acquired 'compensation' at market price was payable. Subsequently the term compensation was replaced by the term 'amount'.

This ideology culminated in the enactment of Urban Land (Ceiling and Regulation) Act 1976 that attempted nationalisation of vacant urban land by paying nominal amount. Finally the fundamental right to property was deleted from the Constitution. The first articulation of the Urban Land Policy was proposed by the Urban Land Policy Committee (Ministry of Health) appointed by the Government of India in 1965. The Committee articulated the following Land Policy Objectives

- To achieve optimum social use of urban land;
- To make land available in adequate quantity, at right time and for reasonable prices to both public authorities and individuals;
- To encourage cooperative community effort and bona fide individual builders in the field of land development, housing and construction;
- To prevent concentration of land ownership in a few private hands and especially to safeguard the interests of the poor and under-privileged sections of the urban society.

Further the Committee observed that to realise the objectives "there is no escape from large scale public acquisition if the question of guiding urban development or the provision of adequate housing and other facilities is to be tackled effectively and large scale advance acquisition of land would really be in the interests of the society as a whole. It is by far the best and perhaps the only way to put an end to speculation in land and to capture subsequent increases in land values. These surpluses, where realised by the public authorities, should benefit the community in more ways than one." Not surprising the role models of Indian Town Planners — Delhi Master Plan, Chandigarh, Gandhinagar and Navi Mumbai were all based on public ownership of land. Whether public ownership in fact achieved the land policy objectives in such cases may be a matter of debate. But a verdict on Delhi experience was;

- It has not been possible for DDA to provide land at affordable prices to low income beneficiaries resulting in large scale jhuggi jhopadi colonies.
- In the absence of price signals land has been sub optimally used, resulting in over provision to powerful groups, and
- DDA's policy to auction very few plots at a time and treating the maximum price quoted in such biding as the real market price has in fact meant artificially increasing the land price through deliberate scarcity."

However, securing large-scale public ownership of land implied compulsory acquisition of land. There was considerable discontent amongst the original landowners about the manner in which compensation was determined and paid. The Land Acquisition Act 1894 initially provided the date of declaration of intention to acquire the land as the reference date for determining the market value. However no time limit was laid down for actual payment of compensation. 1984 amendments introduced the time limit of three years and also provided for payment of interest from the date of award to actual payment or possession of land and solatium of 30 per cent of market value. However the market value is to be reckoned at current use value at the exclusion of expected rise in value on account of future use. The proposed changes in the LA Act and the R and R Policy attempt to remove many of these lacunae.

But planned urban development is not being recognised as a public purpose for which powers of eminent domain could be used and in practical terms the proposed method of deciding compensation and rehabilitation package would make recourse to compulsory acquisition of land expensive for lands that also require substantial investment in trunk infrastructure. This would compel search for new paradigm in respect of urban land. The thinkers in the first world too were enamoured by the socialistic notion of community ownership of land.

However on practical considerations they sought solutions short of nationalisation of urban land. The two extreme proposals had one thing in common. Both considered right to own land and right to develop (or build upon) as separate rights. Henry George argued that a private landowner may have right to own both land and development rights. But he has no rights on the rents accruing to land, as they are results of monopoly and not efforts of the owner. He therefore argued that the state has legitimate right to recover 100 per cent of such rents by way of taxes. He allowed the owner to retain the returns on his investments in improvements. He was also prophetic about the ills of public ownership of land.

In 1879 he stated, "I do not propose to purchase or confiscate private property in land. It is not necessary to confiscate land — only to confiscate rent. Taking rent for public use does not require that the state lease land; that would risk favouritism, collusion, and corruption." On the other hand a committee under the chairmanship of Justice Uthwatt in UK suggested that the betterment occurring on account of development of land should be balanced with the compensation to be paid for acquiring land for public purposes. To enable recovery of betterment the Committee proposed nationalisation of development rights (by paying compensation).

Development required planning permission subject to payment of betterment. Despite three attempts to recover betterment since 1947 it has still not succeeded. A cryptic comment on this reads, "The state expropriates property rights, and then charges those from whom it has taken those rights for granting permission to use them on its terms. A betterment levy is wrong in principle, and like most things that are fundamentally wrong, it will always fail in practice" The Indian urban planning thought under the pre 1991 macroeconomic framework was oblivious of property rights and resultant land and real estate market.

Hernando de Sotto argued the importance of clarity of property rights and labelled poorly recorded property rights as the 'dead capital' unable to ensure finance for poor. But his arguments have evoked little debate in India. Lack of conceptual clarity about land and property rights have given rise to many expedient policy initiatives.

Instrument of Nationalising Development Rights

FAR essentially a zoning tool in US cities rationalises the intensity of development that could be permitted considering the existing level of development, accessibility and use. The FAR in US cities varies considerable from less than one to 15. However in most Indian cities such considerations are not used in defining FAR. In many states common FSI values are prescribed across all cities as part of state wide building regulations and do not form part of master plan of individual city.

There has been two-fold argument justifying this position. First, varying FSI within a city would be seen as discriminatory between different land owners and second, varying FSI across cities would lead to demand from political quarters to increase rationally defined FSI to an arbitrary level proposed in another city. The general tendency has been to prescribe low uniform FSI (around one). This has meant scarcity of development rights particularly in cities that experience faster economic and population growth and resultant increase in demand for per capita floor space. Instead of adopting measures to reduce scarcity of development rights by rationalising FSI pattern, retaining existing low FSI regime is being implicitly used as an instrument of nationalising development rights beyond prescribed FSI (without paying compensation).

Armed with such nationalised development rights the state administrators could allot these rights on conditions of payment or fulfilling other obligations like providing free houses to slum dwellers. (Hyderabad Master Plan proposed a base FSI and permissible increase subject to payment. Maharashtra Government increased FSI in Mumbai subject to payment that would be equally shared between Municipal Corporation and the state government. Extra FSI is similarly allowed in Chennai. Similarly extra FSI is allowed for rehabilitating existing occupants of rent controlled buildings and slums, free of cost). In this context the objectives of raising revenue or helping a class (not necessarily poor) will succeed when base FSI is low and scarcity of development rights is created. Creating scarcity is not a healthy way of managing any market.

FINDINGS AND SUGGESTIONS OF URBAN PLANNING

Urban planning is basically concerned with the location, intensity and amount of land development required for various space using functions of city life industry, wholesaling, business, housing, recreation, education, religious and cultural activities of the people. Thus urban planning is itself neutral. But the institutionalisation of planning practice within a complex bureaucracy has contributed to the re-politicisation of urban planning. It has become a mode of intervention that is only implemented when it serves the specific interests of the interest group parties. Thus slum demolition is a process in urban renewal, a process in urban planning, which has become a mode or tool of the ruling classes for fulfilling their own interests. In shaping the city in such a way that it conforms to the upper class notion of the city, it is its interests, which are furthered. In actuality clearance is necessary to make the centrally located areas available to the capitalists for various economic activities.

As far as Indian urban policies has been concerned it has been seen that there has been a lag between the promises made in different plans in paper and actual practical work that has been done. Though the state within its set structure of the society has tried to work for the poor through different urban planning policies and housing policies like the different slum development policies in Mumbai, urban land (ceiling and regulation) act, more peoples participation in development activities through 74th constitutional amendments, different poverty alleviation programmes but in actual reality very less has been done. The politicians, builders, businessmen, slumlords, the elite class has continuously manipulated the different policies according to there need and profit maximization motivations and left the urban poor in constant misery. Urban policies in general and urban planning in particular has been become an instrument in the hand of the capitalists to fulfil there needs. Thus there is need to make urban development pro-poor or to evolve an urban development framework to agree to the needs of the most vulnerable sections of the urban population.

They should not be concerned about the needs of the capital alone. This can be done through two things. One is the representation in the planning and policy-making bodies and other is the creation of mechanisms or forums for participation in policy making. Thus decentralisation of decisionmaking should be done as has been done through 74th constitutional amendments. But the decentralisation should mean change in the structure or power sharing in the society and not decentralisation of some convenient functions or responsibilities in a centralised society as of now where economic processes and decision making paradigms are centralised. So decentralisation is not only about taking it down to the community but looking at how communities themselves can determine as stakeholders what should happen around them and whatever happens around them should be for them.

People's participation, decentralisation, and Privatisation should not be taken as synonyms of each other. Role of the state and role of private sector and people's needs should be appropriately and adequately discussed so that some general understanding is arrived at, as there cannot be a single model applicable to all the urban housing and basic services and utilities. All dimensions, political as well as economic, of the 74th constitutional amendment be analysed and understood realistically and not on ideological or emotional basis. A proper understanding of urban institutions and their functioning has to be developed. The outcomes of NGO's acting as catalysts in the community actions towards development are very encouraging. The experiences and experiments have been regarding alternative institution building, where the stakeholders themselves are directly brought as actors in the development process.

But NGO's as the solution for housing struggle of the urban poor has also been questioned upon. This has been mainly because their internal limitations set to them like the financial constraints and the external limitations like the power politics and administrative constraints. Housing today is looked upon merely in real estate terms. This is what the real –estate agenda has encouraged today due to the Privatisation thrust in housing and corporatisation of the various development and construction activities. Housing projects are evaluated in terms of size, the built up area, the FSI consumed, the financial turnover, and various other business and marketing merits. The bigger the project, the better it is and the greater the attraction for developers in undertaking the scheme. A huge network is thus established between the developers, the landowners, and the financial institutions wherein the slum dwellers find no place. Thus there is need of changing in the attitude of the government and the elite towards the slum dwellers. Programme for slum development must primarily be seen as an environmental scheme and not merely as an agenda for real estate development and construction turnover. It is the slum-like conditions *i.e.* lack of drinking water, inadequate toilet facilities, garbage, heaps, lack of sewage disposal, absence of open-spaces, inadequate and unsafe access that are of primary concern. Even with the rehabilitation programmes by the Government and the NGO's some basic element is lacking in these projects that is the people. This has been particularly the case of the rehabilitation of the evicted SGNP slum dwellers.

It is obvious from the above argument that neither the state nor the private sector alone can handle the problem of housing for the poor. So there is need for the NGO intervention. But NGO's have there own drawbacks in terms of financial and administrative constraints. So the solution to overcrowding and housing is to ensure that the socioeconomic compulsions that force mass migration towards the cities are ended. However another more rational way out will be the integrated efforts of the private sector, central government, civil society and people. However, this requires simultaneous efforts by national, local governments, civil society, and private sector and people themselves to eliminate impediments at all the levels. While central governments address policy matters and regulatory impediments nationally, local authorities and civil societies should design strategies to make appropriate interventions and regulatory changes in the city. Local experiences should be fed back to national governments to influence their support to cities, as well as for redesigning national programme.

Another thing that can be done is the site and service scheme that has been started by the government in Fourth plan and continued in seventh plan. This is a solution by which the 5 million homeless workers of Mumbai can be housed. In this scheme the government is totally responsible for providing all the services such as water, electricity, sewage, drainage, and toilets, for providing technical know how and skills, interest-free loans and for providing building materials at highly subsidised rates. First of all land is taken over under the Urban Land (Ceiling and Regulation) Act at the rate of Re1/-a square foot which is virtually free or at adequate compensation or if land is already in possession of the government it can be used directly. Instead of This land is divided into thousands of small plots.

Then no construction is actually done in the beginning, as is the case of conventional schemes. But water supply, sewage drainage and electricity are provided to each plot. Then, depending on the financial capacity of each person the worker can build his house. The advantage is even a person with a budget of 2000/-can go ahead. Secondly the house can be developed over time. Thirdly and most important, the worker has control over his house. Housing must be made fundamental right enshrined in the constitution. Both the availability of direct finance and subsidised building materials will tremendously excite the slum dwellers in under taking the renovation and reconstruction of their houses on their own. We can see that people with different income group will be able to afford different types of shelter within there limited finances. Like People with income of as low as ₹.200/-can afford a house of cost 3,200 where the site and services has been provided by the government and the person can make a house within his or her budget. Whereas, a person with a monthly income of more than 1500, can afford a house of ₹.56, 000 made in a conventional way.

Also under the development proposal, the house will be ground and ground + one or two stories high, enabling easy repairs and maintenance directly under the control of the users. There would, therefore, be less dependency on hired skills and services. This will encourage people's participation in decisionmaking and will inculcate a greater sense of belonging resulting in personalisation of spaces and structures. While the construction of houses will be the individual's prerogative, the restructuring of the slum-layout, road, services, open spaces etc., could be a collective effort with governmental support

Development programme for each slum will have to be evolved independently and relevant guidelines fixed. Even F.S.I for each slum may vary to enable housing for all the slum-dwellers on, as–is-where–is basis. Each slum has its own peculiar situation and needs. For example what will apply for Dharavi may not be relevant for another slum in Jogeshwari and vice versa. Therefore, within the main policy framework, individual development strategies will have to be evolved. This will encourage people's participation in decisionmaking.

PROMOTE SUSTAINABLE DEVELOPMENT IN FRINGE AREAS

Control of Development on Fringe Areas: In metropolitan cities and mega cities, urban development is mostly in new settlement areas and new activity centres with planned infrastructures and facilities in the fringe areas to accommodate the increasing population and activities. Unplanned urban sprawl grows around such centres on the agricultural lands taking advantage of the nearby facilities and infrastructure. In such cases, from an environmental perspective, regulations for protecting the agricultural and vacant lands by restricting developments and the stipulations of the regulations.

- No use, other than agriculture or irrigation facilities, is permitted.
- Existing water bodies to be preserved.
- No new building or extension of any existing building exceeding the height of 3.75 metres shall be allowed subject to the total covered area of 50 sq. m.
- The minimum front and side open spaces shall be 2 metres and the minimum rear open space shall be 5.00 metres.

Redevelopment of Blighted Pockets in the Central Area: Blighted pockets of low key commercial areas are observed within the core areas which, due to the small sizes and multiple land ownerships, do not get redeveloped as per the existing regulations. If these areas can be redeveloped in a planned manner much of the demand for commercial floor spaces can be served. For the redevelopment of such areas incentives towards land assembly by amalgamation of plots and additional floorspace ratios need to be given. The regulations in this respect may be:-

- For land assembly exceeding one acre in size, the additional floorspace ratio would be 20 per cent above the permissible limit.
- In case of land assembly of less than one acre the additional floorspace would be 10 per cent above the permissible limit.
- In such cases the developments should confirm the zoning regulations.

Compulsory Rainwater Harvesting in New Area Development: Private developments in the form of sub-division of mother plots (for plotted developments and for apartments) is taking place in the fringe areas and adjoining municipal and non-municipal areas. There are regulations in most of the plans which vary according to the size of the mother plots and the regulations specify the minimum width of roads, the percentage of open spaces, the land for physical and social infrastructures *viz.* drainage, water supply (pump house and water treatment plants), sewerage (sewage treatment plants or oxidation ponds), school, health centre, market, milk booth, post office, power substation etc. In order to comply with the present efforts towards utilisation of natural resources by rainwater harvesting and ground water recharging, the regulations for subdivision should include the mandatory provision of community pools of sufficient size so that the rain water from the area can be stored in such pools. The water may be supplied to the community for uses such as gardening, car washing etc. Adapting physical planning to promote sustainable development: efficient infrastructure planning.

Prescription of street alignments for regional roads in fringe areas: The regional roads (National Highways, State Highways or District Roads), connecting a city with the hinterland are often constricted in the fringe areas due to lack of scope for widening owing to dense developments and abutting built up areas. To avoid such situations where the regional roads pass through vacant areas in the fringe of the city, advance actions for prescription of street alignments may be made without acquiring land. The proposed right-ofway as per future requirements may be prescribed and the regulations.

- For any development on the adjoining plots on the regional roads, the owner/developer would have to make a setback following the proposed right-of-way line and the owners/developers would be allowed the same floorspace as they were eligible for the original plot.

- In such cases after the alignment is notified no subdivision of the adjoining plots would be allowed.
  In cases where the plot size is such that no development is possible by allowing the set back, the local body would have to acquire it.

Dispersal facilities around transport nodes: The railway stations, the regional bus terminals/stops are the important transport nodal points of the urban areas. These areas get congested with dense unplanned commercial and residential developments and in course of time become too congested for easy dispersal of passenger and vehicular traffic.

Given the urban growth rates future public transport will carry larger volumes of passengers. Therefore, the areas required for dispersal facilities will need to be increased. Where such areas are already congested, redevelopment plans are to be prepared by the local body and the redevelopment actions may be initiated, including through private-public partnership projects. In less pressured situations regulations for control of development should be made at least for the area within 200 metres on all sides. In such areas provision should be made for adequate parking facilities for different categories of vehicles including the parking and loading/unloading facilities for transit and paratransit vehicles. Adequate width of the connecting roads, exclusive roads for pedestrians or grade separated pedestrian facilities should be made. In order to achieve these, specific plans should be prepared indicating the future right-of way of the roads, the parking areas for different categories of vehicles, the pedestrian-only roads and the integration with the railway station or the bus terminal. The regulations in such areas may include the following:

- The Floor Area Ratio (FAR) for buildings of different use categories and means of access would be half of the permissible FAR in other areas.
- The compulsory provision of parking spaces for buildings of different use categories would be double of that required in other zones.
- No new cinema halls, theatres and entertainment centres would be allowed within the area.
- For buildings with retail commercial in the ground floor, the minimum front open space would be 5 metres.

Environment protection around relocated hazardous uses: For implementing the development plans, the non-conforming uses located in a scattered way are required to be relocated in the specified areas in the fringe within a stipulated period. Such uses may be the obnoxious and hazardous industries, tanneries etc. Normally, such areas for relocation are selected beyond the city limits and in the vacant and agricultural lands.

When such relocation of activities takes place, unplanned developments adjoining such centres occur and gradually these areas grow. To avoid further environmental hazards, the relocation areas should be provided with a buffer zone where no developments other than agriculture and pisciculture are permitted. In this regard, regulations of the Pollution Control Board are to be followed. But in the development plans there should be specific regulations in respect of the buffer zones. The regulations in this respect may be:-

- The area covered by 200 meters on all sides from the boundary of such areas would be designated as Buffer Zone.
- No developments other than agriculture, pisciculture and plantation of trees would be allowed within this zone.
- The existing residential uses within the Buffer Zone would have to be relocated within a stipulated period.

Adapting physical planning to promote sustainable development: conservation of water resources and waste management

Protection of Water Fronts: The water fronts (sides of rivers, canals, lakes and big ponds) in many cities are encroached by unauthorised users and developed in an unplanned manner. These water fronts need to be protected to ensure proper drainage, and access for open-air recreation, water transportation and protection against soil erosion. Area within 100 metres from the banks should be designated as Water Front zone and specific regulations should be prescribed.

The regulations for such zone may include:-

- No new building within 30 meters from the edge of the banks would be allowed.
- In the area lying between 30 metres and 100 metres from the edge of the banks no building more than 5.00 metres in height and 30 metres along the waterfront would be allowed. In case of buildings on stilts the maximum height of the buildings shall be 6.50 metres.
- There shall be a linear gap of 50 metres between two buildings alongside the water front.

Environment protection around solid waste disposal sites: Solid waste management is one of the most critical problems of cities.

The locations of the intermediate collection sites and the final disposal grounds need special attention in consideration of the environment hazards of the nearby localities. The intermediate collection sites are generally located within or near the settlements and therefore need to have a buffer zone. This buffer zone should cover at least 30 metres on all sides. The regulations in this respect may be:-

- The intermediate collection site may be designated as Inner Disposal Zone.
- The area should be provided with boundary walls of at least 3 metres high on three sides.
- The actual dumping area should be circumscribed by two to three rows of trees in the buffer zone.

POLICIES IN INDIA ON PRODUCTION FORESTRY

NATIONAL AGRICULTURAL POLICY

National Commission on Agriculture (NCA) recommended a change over from the conservation-oriented forestry to more dynamic programme of production forestry. According to the Commission, production of industrial wood would have to be the main reason for the existence of forests and should be project-oriented and economically feasible.

The NCA recommended against leasing forestlands to industry and instead wanted government to allocate them forest raw material through enhancing commercial forestry. For the purpose of commercial forestry, it further recommended that out of 64 million hectare (Mha) of forest lands available in India at that time 48 Mha should be brought under production forestry and remaining 16 Mha for biological diversity.

NATIONAL FOREST POLICY (NFP)

NFP, 1952 acknowledged the need for sustained supply of timber and other forest produce to meet the developmental need of various industries. The policymakers had two major objectives in mind while framing NFP 1952. First to ensure that forests were preserved and managed on a sustainable basis, and second to use them for meeting national interest as a source of timber. Since both these objectives somewhat opposed each other, NFP came up with a concept of '*treelands*'. It identified the scope for involving state governments as well as various other institutions, such as, Defence, Rail-ways, Public Works Departments, Universities and Colleges, Boards, Municipalities and other local authorities, associations and institutions in the process by converting the land at their disposal into treelands. To ensure that the needs of forest-based industries are met without putting undue pressure on forests the need to substitute tree-species of commercial importance in place of inferior tree–species was advocated. The policy also mentioned involving both the industries and individuals in a bigger manner and emphasised on considering commercial and industrial interests for establishing closer contacts and bonding between Forest Research Institutes and industries utilising timber and forest products.

Forest Conservation Act, 1980

The Act put a ban on the de-reservation of forests or use of forestland for non-forest purpose. The Act of 1980, consequently, brought the subject of forests from state list to concurrent list enabling the parliament of India to look into the matter of forests. After the Act, no state government could use forestland for non-forest purposes without prior endorsement by the central government. This put up a check not only on the conversion of forests but also brought to halt the fragmentation of the remaining forests. This ensured that a local party could not take the decision of diverting the forests for taking up any political mileage. The centre acted as an independent agency, to which the state governments and their decisions were answerable.

National Forest Policy, 1988

The NFP, 1988 looked upon forests not as a source of raw material for commercial purposes, but primarily for conserving soil, water and biodiversity besides meeting subsistence requirements of the local people. The policy clearly mentioned that the economic utilities coming from the forests would be secondary to this prime aim.

The NFP 1988 shifted the official focus from fuelwood and timber to the management of forests primarily for their services. It made it imperative to ensure that the forests are not only conserved but also their cover and productivity is increased so as to meet the demand of goods and services coming from forests. The policy, while addressing the production forestry programme, expressed its concern in plugging the increasing demand-supply gap of fuelwood and meeting the national needs.

But it explicitly mentioned that no such activity should result in any sort of clear felling of already existing natural forests. It expressed the desire that such programmes should help the country meet with its objective of achieving one-third forest cover. The policy put a thrust on meeting two important concerns of utilising the wastelands and increasing tree cover by promoting major forestry programmes. To reduce the pressure on forests, it also aimed at promoting substitution of wood besides encouraging better and efficient utilisation of forest produce. Since land plays an important role in taking up plantations, it advocated that a slight modification in land laws should be taken up so that the process becomes smooth for the investors.

At the same time, it maintained that any sort of leasing should be in coherence with the existing land ceiling act. The policy discouraged the earlier trend of meeting raw material requirements of forest-based industry by operating on natural forests. It directed these industries to meet their requirements by establishing a direct relationship with individuals who had the capacity to do so. The industry was encouraged to chip in by contributing its resources and expertise at various levels during the total project duration. The policy also gives more importance to native species over the exotic species. It discourages the use of exotic species without prior scientific trials to establish that there are no adverse effect on the environment.

National Forestry Action Programme, India (NFAP), 1999

The NFAP aimed to prepare an action plan for next 20 years, in conformity with NFP 1988. It is a comprehensive strategic plan to address the issue underlying the major problems of the forestry sector and to reverse the process of degradation for sustainable development of forests. The consumption of fuel-wood in India was reported to be about five times higher than what could be sustainably removed from forests. Further, a large tract of forestland was claimed (nearly 4.3 Mha) for undertaking industrialisation and developing infrastructure necessary for meeting the development goals of the nation. The NFAP aimed at meeting the requirements of forest based industries without compromising the conservation and protection of natural forests.

The basic purpose of NFAP was to establish direct linkage between the NFP and the National Five-Year Plans (FYP). In the past, a comprehensive and constant programme structure for forestry was found missing. Every plan had its own programme structure. So it was difficult to find linkages and establish trends. Although plans had specified objectives and programmes, the main activity under most of them was tree planting. NFAP recognised the capability of plantations to help conserve the natural forests by providing an alternative source for forest products. The Programme was also optimistic regarding the capability of plantations to earn foreign exchange, besides meeting the domestic requirements of the country. The Programme admitted that despite the policy addressing direct relationship between industry and farmers, the government did not adequately support private initiatives. It raised concern on the inability of the government to provide these initiatives with relevant research, extension, technological packages, input delivery, and market information or credit facilities. It felt and remarked that it was imperative to encourage small operators, keep them interested in sustainable forestry development and understand their needs adequately.

NFAP suggested looking at these plantations as a means of raw material for industries or for meeting energy requirements. It intended to conserve and rehabilitate 31 Mha of degraded forests (less than 40 per cent crown density) in addition to bringing 29 Mha of non-forest land under plantations. It advocated the use of different strategies for both these targets. The problem of degraded forests was to be tackled by involving local communities, especially the areas near villages through Joint Forest Management (JFM). For non-forest land, market-oriented massive tree plantation drive involving multiple agencies was proposed. The NFAP came up with a strategic approach to meet its objectives. The approach had five interrelated 'strategic areas' which formed the very basis of the Programme. These approaches were:

- Protect existing forest resources
- Improve forest productivity
- Reduce total demand
- Strengthen policy and institutional framework
- Expand forest area

The state governments while preparing the State Forestry Action Programmes (SFAPs) incorporated all planned activities as part of these activities.The NFAP recommended that for the sustainability of forests, the productivity of forest plantations was to be increased at least 3 to 5 cubic meter per ha per year (m^3/ha/yr) by promoting regeneration and enrichment of plantations. Plantations to be carried out on all categories of wastelands were also suggested. Keeping in mind the fuelwood requirements of the country, emphasis was paid on taking up plantations of fuelwood species on non-forest wasteland. Strengthening the institutions for people's participation in protection and development of degraded and fringe forests was emphasised.

Regarding the financing part of the programme, the sources of funds were grouped under four broad categories, namely, domestic public financing, domestic private financing, external public financing, and external private financing.

NATIONAL ENVIRONMENT POLICY

National Environment Policy (NEP), 2004

NEP emphasised the need to develop a strategy to meet the goal of raising the forest cover of the nation to 33 per cent by 2012. It promoted the involvement of non-forestry sector as well. For increasing the forest and tree cover of the nation, multiple stakeholder partnerships have been recognised where each stakeholder would have his role clearly defined. For achieving the aforesaid target of 33 per cent forest cover, the NEP 2004 suggested that the afforestation activities should be taken up on degraded land, wastelands as well as private land holdings.

Key elements of the strategy would include: (i) implementation of multi-stakeholder partnerships involving the Forest Department, local communities, and investors, with clearly defined obligations and entitlements for each partner, following good governance principles, to derive environmental, livelihood, and financial benefits; and (ii) rationalisation of restrictions on cultivation of forest species outside notified forests, to enable farmers to undertake social and farm forestry where their returns are more favourable than cropping. The policy called for development of a strategy to achieve the targets set for the eleventh five-year plan.

The problem of climate change and its underlying causes in global warming and increased usage of GHGs has been addressed in the Policy. It identifies deforestation in the country as one of the reasons for climate change expresses concern that India and other developing nations would be affected the most by climate change. The current level of emission in India is substantially lower than that of the developed nations; however, India's economic growth could result in an increase in GHG emission. However, government policies favouring renewable energy and afforestation projects as well as the growth of less energy intensive service sectors would result in a decrease in the level of emissions.

PLANNING COMMISSION REPORTS

Report on Leasing of Degraded Forest Lands, 1999

The Report was of the view that the degraded forestland should not be leased out to private entrepreneurs. In accordance with NFP 1988, it recommended that industries needing forest raw material should establish contact with farmers. The government would lease land to Forest Development Corporation (FDC) who, in turn, would enter into proper MoU with the user agency without leasing the land to them, as per the GOI guidelines of 1994. This MOU would give the user agency the right to undertake afforestation and take a fixed percentage of forest produce at the time of harvesting.

The Report recommended leasing barren land far away from habitations, available in plots of 1000 ha or more, and of no use to the villagers. These are desert or ravine or saline lands, which require huge investment before they can be made productive.

The Report gives reference of Investment Promotion Scheme of Ministry of Rural Development, Government of India wherein such enterprises would be entitled to 25 per cent subsidy on their capital investment. This means that the land needed for production purposes could be mobilised for meeting the demand of industry provided the industry is ready to put in huge investment besides following the guidelines. The report estimated that 33 Mha of degraded non-forest lands and 27 Mha of degraded forests (a total of 60 Mha) would be available for tree growing.

The Report also addresses the issue of difference between barren and degraded lands. Uncultivated public land was classified under these two categories. Barren land were defined as having forest/tree cover less than 10 per cent whereas degraded land were classified as having forest/tree cover between 10 to 40 per cent.

Greening India for Livelihood Security and Sustainable Development, 2001

Greening India Programme proposes to cover 43 Mha degraded land (15 Mha of degraded forestland, 10 Mha of degraded irrigated land, and 18 Mha of degraded rainfed land). It recognises that there exist numerous problems that make the implementation of agroforestry programmes difficult:

- Cumbersome legislation with respect to tree felling, wood transportation and processing.
- Lack of market information and infrastructure
- Dearth of appropriate agroforestry models
- Absence of economic security and incentives for tree growers
- Lack of extension training and demonstration
- Non-availability of quality planting material
- Unfavourable Import and Export policy

The Report highlighted the lessons learnt from the farm forestry programme:

- Tree planting should not be undertaken in uncultivable lands, but in cultivated-field and homesteads.
- All wastelands, non-forest areas and degraded forestlands should be brought under silvi-pastoral system with suitable species for fuel and fodder production. This can be addressed through JFM.
- Coastal land through afforestation, ravines and sand dune area for land reclamation, area under mining leases, water–logged areas can be targeted for converting into forested land.

Five-Year Plans

Forestry has also been covered from the first five-year plan, though it remains neglected most of the time in the planning process. Of the total fund allocated to various sectors, funds for forestry in India hovered around 1 per cent only. The expenditure on forestry sector has increased substantially since sixth five-year plan in accordance with the increase in afforestation activities undertaken to meet the desired targets. The sixth five-year plan coincided with the FCA 1980. So the increase in spending on forestry sector can be attributed to a shift in governmental policies. In the fifth plan, the spending was a meagre ₹.107.28 crore, which increased to ₹.15964.06 crore at the end of ninth five-year plan. The area afforested under the same time period has increased from 12.21 hectares to 80.50 hectares. The comparison shows that the spending has increased at a much higher rate than the afforested area. The area afforested under seventh, eight and ninth year plan has been almost same, but the spending has almost tripled in same time period. This shows that growing trees alone will not make these afforestation activities successful. Proper care and maintenance of the area afforested needs higher allocation of money. In fact, money spent in ninth five-year plan was close to 90 per cent of what had been spent till date in the country on these activities. This sends positive signals regarding finance to the project proponents and implementers about the success of afforestation.

Tenth Five-Year Plan (2002-2007)

The current five-year plan targets to increase the forest/tree cover to 25 per cent. It agrees that no strategy would be successful unless and until the basic needs are met. Recognising the role played by JFM in regeneration of degraded forests, the plan recommends taking further steps in identifying its strengths and weaknesses, so that the area under implementation can be increased. Raising concern on high imports of round timber and other forest produce, it expresses a desire for reversing this trend. This could be achieved by utilising community land, degraded forests or private farmlands of the country. Implementing the same would mean removal of government subsidies, regulation of tariff on imports and other such policy modifications, which will make the plantation activities more desirable among farmers and village communities. Promotion of technology, credit support, developing marketing infrastructure and providing extension and training to interested farmers has also been recommended. Policy recommendations for tackling the problems related to constraint of felling, transportation and selling of forest produce which act as disincentive for growing trees are also suggested. This asks for creating a favourable policy environment, which promotes positive interaction among various stakeholders.

Keeping up with the challenges facing the society, plan proposes to utilise the wastelands and degraded lands for taking up CDM projects. This would help in utilising the wastelands besides generating additional incentives for taking up plantations. Emphasis has been put on growing species such as *Jatropha curcas* and *Pongamia pinnata*, which also grow naturally and can be utilised for generating bio-diesel, although success of these schemes are yet to be fully realised.. These plantations are cost-effective and easily replicable. To take up the afforestation activities more seriously all programmes have been merged under a single scheme called 'National Afforestation Programme (NAP)'. It is being operated through Forest Development Agencies (FDAs). Similarly, another programme named National Action Programme to Combat Desertification under UN Convention to Combat Desertification (UNCCD) has been taken up by MoEF. A 20 years' comprehensive NAP to combat desertification in the country was prepared with the following objectives:

- Community based approach to development,
- Activities to improve the quality of life of the local communities,
- Raising awareness,
- Drought management preparedness and mitigation,
- R&D initiatives and interventions which are locally suited,
- Strengthening self-–governance leading to empowerment of local communities.

For the Tenth Five-Year Plan, it has been proposed to initiate activities that include, among others, assessment and mapping of land degradation, drought monitoring and early warning system groups, drought preparedness contingency plans, and on-farm research activities for development of indigenous technology, etc.

The growing demand of raw material from our natural resources is threatening them. Envisaging the threat to the natural resources due to growing demand of raw material,is, the plan Plan proposes to take up plantation, which would reduce pressure on the natural forests and reverse the negative impact of deforestation while meeting the increasing demand. The Plan proposes to encourage agro-forestry by promoting technology, extension, and training, credit support, marketing infrastructure, etc., and providing a policy environment, which assures the farmers of a remunerative price. It also proposes to removes the constraints of felling, transport and marketing of forest produce from private holdings in different States and to formulate a common guideline for this purpose.

The current mean annual increment (MAI) of forest plantations varies from about 2 m^3/ha/year for valuable timber species to about 5-8 m^3/ha/year for eucalyptus and other fast growing species. Generally, MAI is around 10 m^3/ha/year in good quality plantations in various countries. This poor performance of forest plantation remains a great concern in the area of policy decisions. The Plan suggests measures including appropriate site selection, site-species matching, planting of elite clones, proper maintenance and protection, timely tending, thinning, irrigation, application of manures and pesticides, etc. for improving the productivity of plantations. For improving the utilisation of plantation, the plan identifies Indian Plywood Industries Research and Training Institute (IPIRTI) to be actively involved. IPIRTI can also help develop programmes that will not only fulfill the demand, but also provide with more wood substitutes.

POLICIES GOVERNING FOREIGN TRADE IN FORESTRY PRODUCE

Timber price in India has been increasing at the rate of 15 per cent per year, making import of timber more attractive, and Ŧthe current import policy allows duty free import of timber and pulp. In order to help wood-based industries meet their requirements, the government removed the trade barriers and liberalised the import of timber on Open General License (OGL). For industries, it was a welcome step and had an additional benefit of growth of industries in the coastal belt. This policy change led to increased import of timber.

However, the produce from the imported timber is sold mainly in domestic market and there is practically no export of products against import. Export is less than one tenth of import.

India's foreign exchange reserve as well as interest of tree growers has been put under an unwanted pressure. It created a sense of insecurity among tree growers of the country who now had to be satisfied with a smaller share of the market, as global players offered attractive deals to capture a bigger share of the growing market. Despite tough competition from the forestry sector and importing agencies, farmers are still supplying 50 per cent of wood supplies from their holdings. The importance and potential of agroforestry and several other such models has not been realised.

The acceptance of timber from plantations as raw material in wood based industries has opened new avenues for these farmers. However, to give a boost to tree growing activity as a means for import substitution, certain policy initiatives were felt necessary and were recommended.

- To impose heavy import duty on such wood products, which are/can be produced within the country to meet domestic needs.
- Enhance R&D efforts to manufacture quality products from plantation grown wood and other renewable fibres for import substitution.
- Export of wood and natural fibre based products promoted through incentives and simplification of procedure.
- Evolution and implementation of minimum mandatory material and product standards.
- Mechanism for collection and dissemination of information regarding import, export, prices and trade of timber needs to be developed.

The import of various wood-based commodities has been quite high than the exports both in terms of quantity and money involved. Though the exports have increased a bit compared to import in year 1999-00 over 1998-99, huge gap still remains.

Besides, the farmers would find it difficult to make a financial comeback to take up this activity again if their initial investment does not return enough profit. Moreover, import of wood reduces the option of generating additional employment opportunities in the country, which would be easily taken up by the low-income class.

A general reason cited for this import has been to reduce pressure on forests. Production forestry, farm forestry, agroforestry, if utilised to their hilt can do away with this import, which as of now puts a burden of ₹. 8,000 crores annually. Moreover, this would also help achieve the goal of one-third forest cover. Now the industries are also interested in taking up these activities by building relationship with farmers. The north-western part of the country took these efforts on a large scale, and massive plantations of eucalyptus and poplar were raised. However, a lack of organised market along with government policies, which make the process of harvesting trees and their transportation cumbersome, did not allow this effort to get the optimum revenues.

Wood import under Open General License (OGL) made the proposition of importing wood raw material more attractive. Generally, this imported wood comes from natural forests outside the country. In many cases, the harvesting procedure is unsustainable. In a way, these steps save Indian forests at the cost of forests in other wood-exporting countries. In today's scenario of climate change, a phenomenon occurring at one place can have complex implications at far off distances as well. So the problem gets transferred instead of getting solved. So the solution lies in raising more trees within the country. This will provide incentive for local farmers and a disincentive for those countries indulging in these unsustainable activities by taking benefit of imperfect market conditions.

Financing Schemes

Institutional funding is very important in areas where a farmer-industry relationship is to be established. In order to promote the afforestation activities, the central government has been spending money through the five-year plans. Schemes such as Investment Promotional Scheme have been instrumental in initiating and promoting production forestry throughout the nation, though not as successfully as they were initially conceived.

The experience in the last two decades show that institutional funding has been minimal in forestry programmes. These are some important reasons:

- Dearth of technical and economic data on different farm forestry models. This limits the ability of banks to evaluate bank ability of various farm forestry projects.
- Producers and banks often find the associated risk unacceptable. There are no insurance system to guard against the loss to producers and banks arising from various natural calamities.
- The lending banks do not have adequate capability to assist in formulation and appraisal of projects for farm forestry.

Investment Promotional Scheme

The Scheme was launched in the year 1994-95 in order to stimulate involvement of the corporate sector/financial institutions, etc. to pool in resources for development of non-forest wastelands.

The principal objectives of the scheme were:

- To facilitate/attract/channelise/mobilise resources from financial institutions, banks, corporate bodies including user industries and other entrepreneurs for development of wastelands in non-forest areas belonging to Central and State Governments, panchayats, village communities, private farmers, etc.
- To promote group of farmers belonging to different categories, namely, large, small, marginal and SCs/STs for bringing wastelands under productive use.
- To facilitate production and flow of additional biomass including farm-forestry products used as raw material inputs for different types of industries.
- To facilitate employment generation through land development and other allied land based and related activities including plantations.

The Scheme was restructured to make it broad-based and circulated to all the states and other concerned in August 1998. Under this Scheme, Central Promotional Subsidy was limited to ₹. 25 lakhs or 25 per cent of the project cost for on-farm development activities, whichever was less, subject to condition that the promoter's contribution in the project shall not be less than 25 per cent of the project cost. The projects promoted by the Scheduled Commercial

Banks (SCBs), Regional Rural Banks, Land Development Banks and Cooperative Banks were eligible for promotional grant/subsidy under the Scheme. Under the Scheme, 41 projects covering an area of 1435 ha with a total cost of ₹.16.88 crores (firmed up by the bank) and subsidy of ₹. 1.09 crores have been sanctioned up to March, 2004. Because of slow progress, the Scheme was discontinued in 2003-04.

Tax Deductions

Tax deduction is an important financial incentive for production forestry. The Government of India announced tax deduction to companies for carrying out projects of softwood plantation on degraded non-forest land.

POLICY REVIEW AND CHANGE

Government may take up an internal review of the management function relating to collection, processing and marketing of NTFPs, revenue and royalty generated from the trade, institutional arrangements for management etc besides JFM activities in the State. Accordingly, the government would need to make necessary changes in the existing laws and rules.

There should be clarity in defining NTFPs, coherence in the laws and rules, transparent management operations, compliance with National Forest Policy (NFP) 1988 and other national conservation guidelines etc. The Government should develop holistic intervention strategies, and programmes for sustainable management. The Government should develop a database on the NTFPs, and the administrative reports of the government should clearly reflect the status of NTFPs in the State.

PRICING OF NTFPS

State level price fixation for the NTFPs is a difficult proposition. It attracts criticism from various quarters. It is always better for the Government to accommodate variations in prices in different geographical regions in order to take care of the local demand and supply considerations, transport, storage etc. At the State level, a floor price could be fixed, and variations may be allowed at the district level, beyond the floor price.

By this, the forest dwellers could appropriate the benefit of price advantage. District level price fixation committees could be set up under the chairmanship of the District Collector with DFO as the member secretary. Other members could include the District Horticulture Officer, the District Industries Officer, one representative each of the TDCC and OFDC, two representatives from the local NGOs and a representative from the local industries. There should be at least 30 per cent representation from the primary collectors to include both tribal and non-tribal members (at least one of whom should be a woman) representing all the areas from the district. The total strength of the committee should not exceed 12.

The committee could meet much before the commencement of NTFPs leasing year *i.e.* September. Instead of having one meeting in a year, it could have at least one meeting in a quarter. Before fixing up minimum support price for the NTFPs for the year, the committee could institute a review on the status of NTFPs in the district, and the problems in marketing network, and the findings of the same could be used to understand the problems and develop strategy to address these by the committee. The minimum support price could be fixed, based on the principle of incremental margin, working backwards from the actual market price. It should also take into consideration the prevailing prices in the bordering States. In Andhra Pradesh, the Government through Girijan Co-operative Corporation follows the same procedure for fixing up prices for the NTFPs.

LEASING OF NTFPS

The Government should gradually discontinue the existing practice of monopoly leases to Joint Sector Company, private parties, Paper Mills, and also to the Government Undertakings. Accordingly, once the royalty, minimum support price etc are properly decided upon and regulated, leases could be given to a number of buyers including co-operatives, non-profit making societies, VFCs, VSSs and their federations etc on competitive basis for ensuring maximum procurement and fair price to primary gatherers. In the areas, where CFM/JFM has been in vogue, the responsibility of primary collection, storage, minor processing etc of the forest produce could be entrusted to the JFM institutions. In the areas where CFM/JFM has been in vogue, the responsibility of primary collection, storage, the responsibility of primary collection minor processing etc of the forest produce could be entrusted to the committees.

ROYALTY

Royalty provision in the Forest Acts is a major source of revenue from NTFPs (Kendu leaves, Sal seeds, etc.) which is either paid by private traders or state agencies. The unsustainable commercial extraction of NTFPs in many cases has led to destruction of precious plant species and the consequent environmental imbalances. The Government could constitute a high level committee for preparing guidelines for royalty fixation of different NTFPs and also to review the situation notwithstanding the welfare interests of the primary gatherers. For this, collection of NTFPs, minor level processing etc could be taken up by tribal cooperatives, VSSs and VFCs. However, the royalty has to be reasonably low in order to promote more and more local institutions in managing the NTFPs.

PROCUREMENT OF NTFPS

The departmental agencies and other leaseholders could open collection centres at least one each in a Gram Panchayat area. A copy of minimum support price fixed by the district price fixation committee, issued by the District Collector could be displayed at the collection centre. The price list of NTFPs fixed by the district committee needs to be circulated by the DFO as well as the Assistant Registrar, Co-operatives among the lower level forest officials, lessees, VFCs/VSSs and other forest protecting communities, NGOs, Gram Panchayats etc. Where VSSs have been constituted and traditional CFM groups exist, more freedom should be provided to them for collection, storage and processing of selected NTFPs.

A joint committee of the forest officials and representatives of primary collectors needs to be constituted at the Forest Range Level to monitor the procurement of NTFPs including the payment of minimum support price by the lessees. The Government could plan for setting up permanent collection centres at the Forest Range level along with necessary reserve fund for collection of NTFPs.

MARKETING OF NTFPS

Some degree of freedom needs to be given to the primary collectors and producers to sell NTFPs collected by them in the local markets for the consumption by the local population. Trading restrictions need to be relaxed, so that the local people can store NTFPs for longer periods and be provided excise licenses for primary level trading. This needs changes in the forest laws and rules.

Market network information system needs to be established and monitored regularly. Information pertaining to product profiling, product development, prices, market trends, finance etc. need to be generated on a regular basis and disseminated to the target groups. All this and other relevant information should be disseminated regularly in the meetings of the VFCs/VSSs and Panchayats, and by the field-level extension workers of different Departments.

Market Promotional Boards (MPB) may be set up at the district level to provide information to the stakeholders regularly, and also to establish linkages with different trading houses to market the produces of the area. There is a need for setting up of a State-level apex institution devoted exclusively for the development and management of NTFPs in Orissa. Research and development, capacity building of the local level trading institutions, community institutions involved in processing and marketing of NTFPs, departmental agencies etc., linkages with the industrial houses, lobby with the exporting agencies, organising finances from the financial institutions for processing and marketing both at State and micro level etc. should be part of its functions.

INDIGENOUS KNOWLEDGE

The current scale of operation of NTFPs collection and the income from sale indicate the potential of forest resources in the JFM areas of the State. The fact that bulk of the products is sold in raw form is a pointer to the vast potential for processing of forest products and the benefits of value addition at the household level. In this respect, it is important that the indigenous skills, knowledge and experiences of the forestdwellers gained over years are fully utilised for maximising benefits. An interface between traditional knowledge and modern concepts needs to be forged for NTFP production, marketing and processing with some amount of value addition. Such a policy strategy could enhance the socio-economic capabilities of the forest-dependent poor in a big way to secure food security on sustainable basis.

PREVENTION OF ILLICIT TRADE

There is a need to make competitive price available to primary collectors to plug leakages in illicit sales of precious NTFPs resulting in unscrupulous trade practices. Involvement of the people at the grassroots level would be most crucial. Competition among the private traders/businessmen, State Agencies would not only ensure price advantage, but also maximise procurement of NTFPs in the region. Accordingly, forest policies and provisions in the Acts need to be changed and reoriented followed by periodical market surveys as well as market intelligence network to ensure people friendly results.

SOCIAL PARTICIPATION AND FOREST POLICY

Decentralization in Bolivia was motivated by increasing pressure from civic committees seeking to have greater control over natural resources, the general trend towards decentralization in neighbouring countries and the prominence of decentralization in donors' agendas. Its implementation has been affected by social participation dynamics, and the new forestry regulations have had implications for democratization of forest resources access.

ANTECEDENTS OF DECENTRALIZATION

Bolivia has three levels of government: central government; departmental government or prefecture, whose main authority (the *prefecto*) is appointed by the president; and municipal government, in which democratically elected municipal councils elect the mayor. Before decentralization most decisions were made at the central level. The municipalities had limited resources and little influence in policy-related decisions, even those directly affecting the development of their municipal jurisdictions.

The Bolivian lowland forest area was largely marginalized from the political centre in La Paz, where most political decisions were made. In the early 1960s that region was progressively integrated into the national economy through the expansion of natural gas extraction, agriculture and logging. Those factors fostered the growth of a regional elite, expanded the contribution of that region to the national income and increased its influence on development programmes. In the 1970s, the government took its first step towards decentralization by establishing corporations for regional development. This was closer to an administrative attempt to transfer some investment decisions to the departmental level but the central government still appointed the presidents of such entities, and most of the decisions were negotiated at higher levels. Civic committees (groups of local social organisations) had been seeking greater access to forest revenues and greater participation in the formulation of forest policies since the mid-1970s. In the 1980s, legislation was approved establishing the collection of a forest fee of 11 per cent, to be used in regional development projects. (In 1993, further legislation was passed stating that forest companies would pay 80 per cent of their taxes directly in the areas where the resources originated, but the mechanism was hard to implement in practice.) By the late 1980s, the national forestry service was deconcentrated to local branches. This move towards deconcentration, however, did not make the forestry service a more efficient and less corrupt institution. Municipal authorities, meanwhile, continued to respond to the leaders of their respective political parties rather than to their constituents.

Decentralization was prompted through approval of Popular Participation Law (No. 1551) and Administrative Decentralization Law (No. 1654), both passed by Congress in 1994. The first altered the responsibilities of the municipal governments, while the second modified the responsibilities of prefectures or departmental governments. In 1996 a new Agrarian Reform Law was issued, as well as a new Forestry Law. The Agrarian Reform Law's objective was to define the legal basis for a system of titling and land regularization, and to redefine the conditions of access to rural property. The Forestry Law attempted to redefine the conditions for obtaining and maintaining forest rights. Both affected the way in which landholders and forest users can access and maintain their rights for forest resources use.

PARTICIPATION IN INVESTMENT DECISIONS

The Popular Participation Law expanded the municipal government's jurisdiction beyond the urban centers to the whole rural area within the municipal borders. It made municipalities responsible for local schools, health facilities, roads maintenance and water systems. To finance these new responsibilities, the central government allocated 20 per cent of the national budget to the municipal governments, to be distributed among municipalities in proportion to their populations. Both rural and urban property taxes were earmarked for the municipal governments, who now administer their collection. The Popular Participation Law has strengthened municipal governments and made them more democratic. Rural populations – mainly smallholders and indigenous people – have gained the right to participate in municipal elections and run for the municipal councils. Nonetheless, national political parties still appoint individual candidates, allowing political parties to maintain their control over local political agendas, and to reproduce a system of political patronage with local leaders. Local candidates who want to run for office have to negotiate their agendas within the priorities of the political parties, and if they are elected they are accountable to those parties.

Furthermore, the law sought to introduce community control over municipal governments by local social organisations (ie local farmer organisations, neighbourhood committees and indigenous groups) and community-based vigilance committees. Nevertheless, these committees' representatives are exposed to continuous pressures from political parties.

Rules and Regulations for Forest Use

The Forestry Law of 1996 defined a set of regulations for forest use, somewhat differentiated according to forest user, under the premise that sustainable forest management is feasible under the right practices. It established a new system for monitoring forest management, enforcement and sanctions to illegal logging, as well as introducing some market-oriented regulations and taxes to discourage unsustainable forestry operations. The goal is to achieve sustainability of forest management through progressive incorporation of less valuable timber species and the application of extraction techniques that promote natural regeneration. Furthermore, the law seeks to create clear rights over forest resources, thereby encouraging investments in forest management; and to eliminate forest crime and illegal logging, as well as set technical criteria for forest management. The public institutional system of the forest sector was substantially altered. The Ministry of Sustainable Development and Planning is the ruling entity, the Forestry Superintendence is the regulation entity, and the National Forestry Development Fund is the financial entity.

Non-commercial forest uses do not require authorization, and a forest management plan is an essential requirement for all types of commercial forest activities. Hence, forest concessionaires as well as private landholders are obligated to design management plans as an instrument to regulate commercial logging activities, including forest inventories and mapping. Forest management plans have to comply with many technical requirements. Forest management, when based on selective management, must respect a minimum cycle of 20 years between logging operations on the same area, and a minimum cut diameter must be respected. Furthermore, annual operations plans are required. The regime to be applied to non-timber forest products is similar.

Democratizing Access to Forest Resources

The new forestry regulations included two provisions that have had some impact in democratizing access to forest resources. The first refers to the exclusive right of indigenous peoples to use the forest resources within their indigenous territories, recognised legally by the Agrarian Reform Law. According to this law, indigenous claims are considered titled after completion of a process of land regularisation. Currently, a total of 19 million ha have been claimed for titling as indigenous territories, but it is not known how much will ultimately be effectively titled in this way. Stocks (1999) estimated 5 million ha to be the area with commercial logging potential in areas claimed as indigenous territories. A provision of the Forestry Law states that local forest user groups can benefit from forest concessions within areas declared as municipal forest reserves, which represent up to 20 per cent of public forest within each municipal jurisdiction. Local forest users can be granted a forest concession if they are recognised as a local user association by the Ministry of Sustainable Development and Planning. This mechanism was conceived as a way to formalize the access for local forest users or small-scale loggers who, because they had no legal right to access forest resources, were previously conducting forestry operations informally.

POWERS TRANSFERRED TO MUNICIPAL GOVERNMENTS

According to Ribot (2001), outcomes of decentralization depend on the type of powers that are transferred to lower levels and on who receives such powers. In Bolivia, as well as in most countries of Latin America, decentralization has followed a top-down format, and municipalities have been the main recipients of authority transferred from the central level.

A Top-Down Model of Delegation

Although the Popular Participation Law did not grant municipal governments any new explicit function related to natural resources management, it motivated some municipal governments, as a result of their larger political authority, to become involved in natural resources issues and thereby capture part of the benefits. Because mayors became relatively more powerful, the central government and the international donors began to consider them more seriously as partners in environmental projects. The increasing political power of municipalities led the government to consider them for dealing with some problematic issues, such as social monitoring of illegal logging and formalizing forest rights for small-scale loggers and other local users. These were included in the Forestry Law of 1996.

The current powers and functions are divided as follows:

- formulate forest policies, strategies and norms;
- determine land classification and evaluate forest potential;
- prepare demarcation of concession areas for timber companies and local groups;

- set prices for concession fees and volume-based taxes;
- promote research, extension and education; and
- solicit technical assistance and funding for forestry projects.

Forest superintendence (SF):

- supervise technical compliance with the forestry regime;
- grant management rights to eligible forest users;
- approve forest management plans for different forest rights;
- enforce forest regulations and sanction illegal forest users;
- issue concessions, authorizations and logging permits;
- request external forest audits of forest operations; and
- collect concession fees and volume-based taxes and distribute them.

Prefectures:

- formulate forest development departmental plans;
- develop forest research and extension programmes;
- promote programmes of rehabilitation of degraded forest systems; and
- develop programmes for strengthening municipal forestry units' institutional capacities.

Municipal governments:

- propose the delimitation of municipal forest reserves up to 20 per cent of available public forest;
- protect and conserve the reserve areas until they are conceded to local user associations;
- inspect and control all forest activities within their territorial jurisdiction;
- report violations of forest regulations to the Forestry Superintendence;
- provide support to local forest users in implementing their management plans;
- establish the registry of forest resources in their jurisdiction;
- develop soil use plans corresponding to the departmental use plans;
- organize training events for local forest users; and
- facilitate and promote social participation in local forest development.

To carry out their new responsibilities, municipal governments are expected to create municipal forestry units. Municipalities can form consortiums with other municipalities to create such units. In theory, the entire system should be entirely financed with the revenues coming from both concession and clearcutting fees.

Prefectures receive 35 per cent of the concession fees and 25 per cent of the fees charged for clearcutting operations. Municipal governments get 25 per cent of both types of fees. The National Forestry Development Fund receives 10 per cent of the concession fees and 50 per cent of the clearcutting fees. SF gets 30 per cent of the concession fees. A reduction of forest concession fees in March 2003 has diminished the financial resources going to prefectures and municipalities.

Restricted Powers for Municipalities

Municipal governments must comply with national regulations regarding property regimes and forest regulations. Hence, their autonomy to make decisions about natural resources depends largely on decisions made at the central level. The legislation thus regards municipal officials as rule followers rather than rule makers, and as implementing agencies for policies defined at the central level. In short, the new institutional forest system has not delegated important responsibilities to municipalities and has therefore not led to dramatic changes. The central level reserves for itself major decisions regarding allocation of forest resources rights, granting forest concessions, approving forest use regulations, and collecting taxes from forest resources use. Municipalities have little to say in these areas, and all informal actions of municipalities for collecting forest taxes and controlling timber transit are considered illegal.

Although municipal governments are empowered to control forest crime, promote community forestry and take other forest-related actions, their power to make autonomous decisions regarding the allocation and use of forest resources is restricted. For instance, they can decide how to allocate the forest resources within the municipal reserves but cannot decide about the size of such reserves. This reflects the tension between the central level, which defends the strong role of the national forestry service, and those who support a more active and autonomous role for municipal governments.

Limited Autonomy for Local Forest Users

The situation of indigenous groups has improved. Indigenous communities whose forest resources were subject to encroachment from illegal loggers now have exclusive access to resources within their territories, though problems persist. However, before indigenous people can take commercial advantage of forest resources, they have to develop forest management plans according to Forest Superintendence regulations, and they have little scope to adapt such norms to their own management practices.

And even though specific regulations for forest management in indigenous areas have been approved, local knowledge has been ignored in much of the forest policy. Enforcement of indigenous property rights depends on some institutional arrangements made at the departmental and regional levels, over which indigenous people have little influence. This issue is linked to a more complex bureaucratic process of land use planning and titling with poor overall outcomes.

The other social actors who have benefited from access to forest resources – small-scale timber extractors and other local forest users – also have no autonomy to make decisions about the way in which they use the resources, and all of them have to comply with forestry regulations to maintain access to public forest resources through forest concession systems.

The Forest Superintendence has instituted command-and-control mechanisms to enforce the implementation of what it considers good forest management practices among local forest user groups and small landholders. Those practices, though appropriate for large-scale forest concessionaires, act as barriers for small-scale forest users.

Implementation of Decentralisation Policies

There are always failures in the implementation of policies, and decentralisation is no exception. In Bolivia, transferring responsibilities to municipalities meant building local capacities in municipal governments, as well as interacting with actors who had disparate incentives and interests to use the forest within the new institutional context.

Building Institutional Capacities

Municipalities with forest resources began to receive their shares of forest taxes in 1997, when they started to set up municipal forestry units. The resources transferred to municipalities have tended to decrease because forest concessionaires did not comply with the forest fees payments. In 2002, a new system formally reduced the amount of taxes collected by the state from forest operations to almost half of what was originally expected, formalizing the reduction of transfers from the central government to municipalities. By the end of 2001, almost all municipalities with forest resources (about 109) had created their municipal forestry units, and each had at least one forest or agricultural technician.

In municipalities with difficult access and little population, the forest units are the only local providers of technical services, and governments tend to value their role in local planning. The main limitations are the staffs' lack of technical skills and inadequate training in social issues, such as conflict resolution.

Building technical capacities in municipal forestry units constituted an important step forward in relation to the past. Nevertheless, progress is not uniform and varies with the financial support the units receive from the municipalities.

That in turn depends on the amount collected in forest fees and what proportion of it goes to supporting operational budgets. A significant portion of municipalities consider it enough to provide minimum resources to the municipal forestry units and spend the rest, if any, on other activities. It is worth mentioning that local government priorities are providing social services and infrastructure, and they have little motivation to support productive projects, though this is changing.

Incentives for the Actors

The new institutional system resulting from both decentralisation and the shift in forest policies has modified the incentives for municipalities and forest stakeholders to continue doing what they were used to doing, or to adapt their social and financial strategies to the new conditions. The main incentive for municipalities to engage with the new decentralized order of things was the likelihood of getting a share of the forest taxes, as well as the possibility of administering the municipal forest reserves. Municipalities also are interested in penalizing illegal clearcutting due to the fact that they get a portion of the fines, but this does not happen in relation to illegal logging. Furthermore, municipal governments have been interested in being active players on forest issues in the cases where local forest users have political influence on local decisionmaking, or in cases where forest-dependent people represent important votes to keep them in office.

The incentives of local forest users to engage in the process are diverse. Indigenous people have received important benefits from forest policy reform and decentralisation. They have gained rights to make exclusive use of their forest resources within their territories, and have the chance to expand their influence to participate in municipal decision making. In this context, forest management might increase the social legitimacy of indigenous people's claims to land.

In turn, local small-scale timber extractors' main incentive has been to get formal access to forest resources through the forest concession system, and to benefit more from formal markets. Furthermore, this group should benefit from the technical assistance provided by the municipal forestry units, which are receiving resources from the forest taxes to support such activities. Both groups have benefited from several forest projects (eg BOLFOR) and NGOs with an interest in supporting community forestry.

In the short run, the main losers under decentralisation were absentee forest concessionaires, who must now negotiate with the forestry service. Furthermore, a forest tax has been imposed, and they have had to acknowledge indigenous people's demands. The main incentive for forest concessionaires to adapt to the new conditions was their interest in keeping their forest areas, and an implicit commitment from the central government to help them through a difficult financial situation originating in the timber crisis in regional markets. Municipal governments reacted in contradictory ways regarding forest concessions. Although some are critical of concessions' activities within their jurisdictions, others consider them sources of economic growth and employment.

Homogeneous Solutions for Heterogeneous Municipalities

The Bolivian model of decentralisation does not account for regional variation, and its design assumes that all municipal governments will react uniformly to the challenges arising from the new conditions. Reality proved rather different. Implementation of decentralisation has had to face three issues: uneven distribution of resources to municipalities, the minimal availability of public forest declared as municipal forest reserves, and the hetereogeneous interests of municipalities regarding forest-related activities.

Uneven distribution and allocation of financial resources. The financial resources allocated by the Forestry Superintendence to municipalities have varied. From 1997 to 1999, only 30 municipalities benefited, with more than 80 per cent of the total resources transferred to them. In the richer municipalities, not all income from forest taxes is spent on forest-related activities; some is diverted to other sectors. The poorest municipalities likewise have urgent demands that are outside the forestry sector.

Minimal municipal forest reserves. Even though the law specified that up to 20 per cent of public forest would be declared municipal forest reserves, in practice such areas were not available in all municipalities – in some cases because of the existence of overlapping claims over public forest, in others because public lands were maintained as forest concessions. In some municipalities there were no areas to declare as municipal reserves to allocate as concessions to small-scale loggers. By early 2001, from a total of 2.4 million ha demanded by municipalities as municipal reserves, only 681,000 ha had been designated.

Municipalities' priorities. Only a few municipal forestry units have accomplished the functions they have been granted. Municipalities prioritize their investment in response to social pressures or to local authorities' political motivations. The level of municipal engagement is also related to municipalities' incentives to respond to their constituents and be accountable to higher authorities. For instance, whereas some municipalities are interested in allocating resources to establish a system of forest concessions within municipal reserves, others choose to control illegal clearcutting or make forest management viable for small farmers.

Uneven Outcomes

Transferring responsibilities to municipalities has modified the political and institutional arrangements for forest management, with implications for both people and forests.

Participation in Local Politics

In many lowland municipalities, small farmers, indigenous people and small-scale loggers have been elected to office for the first time. Indigenous groups have been able to obtain political support from the municipal councils to reinforce their land claims, small-scale loggers have obtained support to negotiate temporary logging authorisations, and small farmers have obtained support favouring their efforts to modify land-use and forest regulations. Thus, municipal governments have contributed to political support for some local actors' claims on resources. This is also the case where these groups have strong organisations that can influence the municipality's decisions or where they represent the majority of voters. In these cases, municipal governments may amplify the demands of social actors. In other cases, however, transferring responsibilities and resources to municipalities has reinforced the power of pre-existing local elites – elite capture, particularly in municipalities of northern Bolivia, and where cattle ranchers and timber companies are highly influential in local politics. These local elites have, in some cases, influenced municipal governments to build local alliances against indigenous land claims and strengthened their power over use of resources. The local elites of these municipalities often benefit from activities based on mining of natural resources and promote concentration of benefits in a few hands. Decentralisation, in these cases, tends to produce undesired results in social equity and forest conservation.

Where the social composition of local government is more complex, local elites have to negotiate with small farmers and indigenous people. It is not unusual to find alliances built among the different social groups that support certain development agendas, such as infrastructure development and basic social services. Those social agreements are, however, difficult to arrange for natural resources management, though in some cases alliances have been established to protect conservation areas against encroachment of foreign timber and mining companies (ie municipalities of Rurrenabaque and San Ignacio de Velasco).

Much of the impact of decentralisation on reconfiguring social participation in local politics depends on the local political economy and on the social capital of marginalized groups. In other words, the more democratic municipalities have traditionally had more equitable access to resources. Those in which local elites have imposed their interests are affected by the extent to which local groups have been able to build social capital.

RIGHTS OF FOREST-DEPENDENT PEOPLE

One of the most striking features of decentralisation in Bolivia is that the new, more diverse social composition of municipal governments has had little effect on how forest rights are allocated and to whom: such decisions remain at the central level. In this regard, municipal governments become institutions for stimulating negotiations among local actors and representing local demands to the national level. Indeed, municipal governments were active players in making some forest regulations more flexible, which benefited small-scale loggers and small landholders. The reinforcement of some property rights that accompanied decentralisation, such as indigenous territories and municipal forest reserves, had significant implications for local forest users, though the pace of such changes was extremely slow. Titling of indigenous territories is a slow, bureaucratic process, and indigenous people cannot benefit fully from their forest resources because they do not fulfill the conditions to formulate and implement forest management plans.

Forest concessions for small-scale loggers have also proceeded slowly. Identification and demarcation of municipal forest reserves by the municipalities was relatively quick, but the approval of such areas by the Ministry of Sustainable Development and Planning was slow. Besides bureaucratic procedures, one of the factors limiting the formal creation of municipal reserves was an extremely slow process identifying public forest, which is a precondition for establishing such reserves.

Many of the claimed areas are not available. Moreover, about 26 demands of local user associations were not even processed, leaving the door open for such groups to persist in illegal logging activities. The impact of municipal forest reserves is still small: only 13 groups with 387,000 ha had an approved forest management plan by early 2003. Unfortunately, there are no assessments of the extent to which such groups have improved their incomes from their logging operations. Members of the local user associations, lacking any tradition of collaboration because they were accustomed to working independently, have faced internal conflicts of leadership (Cronkleton and Albornoz, 2004). Furthermore, these groups' organisational procedures and mechanisms for distribution of profits from logging operations remain unclear. Building social capital in these groups might be more important in the long term than improving their practices of forest management.

Impacts on Forest Crime

Many changes in forest management practices can be related to the shift of forest policy. Indeed, the Forestry Superintendence has been more active in monitoring forest crime, both illegal logging and clearcutting, though it was not able to slow illegal activities because of its limited institutional capacity to enforce forestry regulations. There are no reliable estimates of illegal logging, and those of clearcutting suggest that illegal deforestation would be around 80 per cent of total deforestation. At first, monitoring activities focused on enforcing small landholders' and illegal forest users' compliance with the forest regulations. Thereafter they focused on municipalities with high rates of informal activities. Although the Forestry Superintendence carried out most monitoring activities after decentralisation, gradually it sought help from the municipal forestry units – both to use the resources transferred to municipalities to develop monitoring activities, and to justify its actions before the local population. Nevertheless, municipalities' response to forest crime monitoring was ambiguous. Controlling illegal forest activities would affect some politically influential people within some municipal governments, and units had no capacity or resources to spend on activities with little financial return and high political cost. That institutional behaviour has created ambiguous signals about the role that municipalities play in forest crime monitoring.

Municipal governments have more incentive to control the operations of large-scale forest concessions and control illegal clearcutting because of the direct benefits they obtain. A few municipalities have taken over machinery of timber companies and intervened when forest concessionaires logged in areas outside their boundaries. Local authorities complain about the system of auctioning confiscated illegal timber because they do not benefit from it. They have been active in temporarily resolving uncertainty over property rights, primarily of small landholders, by issuing land 'possession certificates', though these lack legal value.

Institutional Tensions and Collaboration

The municipalities and the Forest Superintendence are beginning to build partnerships to improve forest management monitoring at the local level. This agency was initially skeptical about local authorities' ability to intervene in forest-related issues in a politically neutral way. The municipal units view it as undermining the livelihoods of local forest users who have undergone severe difficulties to adapt their practices to the new forestry regulations. That has changed to some degree, and the Forest Superintendence is now more involved in building collaborative agreements with municipal forestry units for regulations enforcement. The driving principle is the need to enhance local operational capacity and make legitimate decisions at the local level. Through formal agreements of collaboration, the agency delegates some responsibilities to certain municipal forestry units. In other cases, however, there is a lack of communication and trust among the personnel involved and some previous negative experiences.

NGOs have helped municipal forestry units overcome institutional deficiencies and build capacity for monitoring, planning and proposal elaboration. Some forest projects – like the BOLFOR project working with local user associations, and the UNDCP-FAO project working with colonists – attempted to build capacity for forest management through training in preparing and implementing forest management plans, developing market skills and in some cases establishing partnerships with timber companies. In most cases, however, forest projects have not included municipal governments working directly with forest users. That might have had some positive effects in such projects' achievement of their goals, but inclusion of municipal units would have contributed better to the effort to develop technical skills within municipalities.

The missing link in the system is at the level of the departmental governments. Even though they receive resources from a portion of the forest taxes, they have resisted spending them on forest research and extension and strengthening municipal forestry units. Instead, those resources were diverted to cover other expenses. An exception to highlight is an isolated project aimed at strengthening municipal forestry units developed by the Prefecture of Santa Cruz between 1998 and 2000, which has been documented elsewhere. There is an ongoing debate over the role of prefectures within the forest institutional system.

Bibliography

A.E. Osmaston: *A Forest Flora for Kumaon*, BSMPS, Dehradun, 1994.

Ajay S. Rawat: *Forest Management in Kumaon Himalaya*: *Struggle of the Marginalized People*, Indus Publications, Jaipur, 1999.

Ajay Sharma: *A Text Book of Forest Trees Entomology*, International Book Distributers, Delhi, 2007.

B Venkatesh; B K Purandara and K S Ramasastri: *Forest Hydrology*, Capital Publications, Delhi, 2007.

Barrows, M.: *A Survey of the Intestinal Parasites of the Primates in Budongo Forest*, Uganda. Glasgow University, 1996.

Benu Singh : *A Survey of the Forestry Research*, Vista International Publication, Delhi, 2011.

Benu Singh: *A Modern Book on Forestry and Horticulture*, Vista International Publications, Delhi, 2010.

Chakrabarty, Falguni: *Adaptation of the Santals to the Hill-Forest Environment*, APH, Delhi, 2012.

Christian Gonner : *A Forest Tribe of Borneo : Resource use Among the Dayak Benuaq*, D.K. Printworld, Delhi, 2002.

Deepak Sharma and Harpreet Singh : *Guinea Fowl : Genetics and Breeding*, Satish Serial Publishing House, Delhi, 2013.

Falguni Chakrabarty: *Adaptation of the Santals to the Hill-Forest Environment*, APH Publications, Delhi, 2012.

G P D Vyas: *Community Forestry*, Agrobios Publications, Jodhpur, 2006.

Gerard Bodeker, K.K.S. Bhat, Jeffrey Burley and Paul Vantomme: *Medicinal Plants for Forest Conservation and Health Care*, Daya Publications, Delhi, 2005.

Gilbert Wooding Robinson: *Soils : Their Origin Constitution and Classification*, Biotech Books, Delhi, 2005.

Gupta, O.P. : *Water in Relation to Soils and Plants : With Special Reference to Agriculture*, Agrobios, Delhi, 2002.

Herminie Broedel Kitchen: *Soils and Crops : Diagnostic Techniques*, Satish Serial Publishing, Allahabad, 2004.

Hutchinson, R. H. S.: *Chittagong Hill Tracts,* Vivek Publishing House, Delhi, 1978.

K.M. Bhat, K.K.N. Nair, K.V. Bhat, E.M. Muralidharan and J.K. Sharma: *Quality Timber Products of Teak from Sustainable Forest Management*, Kerala Forest Research Institute, 2005.

Klaus Seeland and Franz Schmithusen: *Indigenous Knowledge, Forest Management and Forest Policy in South Asia (Proceedings of an International Seminar held in Kathmandu, Nepal in 1998)*, D K Printworld, Delhi, 2003.

L K Jha and P K Sen Sarma : *A Manual of Forestry Extension Education*, A.P.H Publication, Delhi, 2008.

M. Follower : *Genetics and Genetic Engineering*, Rajat Publication, Delhi, 2003.

M.P. Singh and Sunil Kumar : *Genetics and Plant Breeding, Vol. I and II*, APH Publication, Delhi, 2009.

N. L. Bor: *A Manual of Indian Forest Botany*, Asiatic Publications, Delhi, 2010.

N.H. Ravindranath and P. Sudha: *Joint Forest Management in India: Spread, Performance and Impact*, Universities Press, 2004.

Nandini Sundar, Roger Jeffery and Neil Thin: *Branching Out: Joint Forest Management in India*, Oxford University Press, Jaipur, 2001.

Prasad, T.V.S. : *Soil Chemistry : Nutrient and Water Management in Agricultural Soils*, Dominant Pub, Delhi, 2009.

Ram Prakash: *Forest Management*, International Book Distributers, Delhi, 2006.

Reynolds, V.: *Budongo: A Forest and its Chimpanzees*, London, Methuen, 1965.

Santhi,R and K M Sellamuthu: *Fundamentals of Forest Soils*, Satish Serial Pub, Delhi, 2008.

Sathe, T.V.: *A Textbook of Forest Entomology*, Daya, Delhi, 2009.

Shiel, D.: *The Ecology of Long-term Change in a Ugandan Rainforest*. D. Phil., Oxford University, 1997.

Somani, L L and P C Kanthaliya: *Soils and Fertilisers at a Glance*, Agrotech, Delhi, 2004.

Sudharmai Devi,C.R. : *Analytical Procedures in Soil Science and Agricultural Chemistry*, Agrotech, Delhi, 2004.

Taank, Praveen: *Advances in Forestry Research in India*, Cyber Tech Pub, Delhi, 2009.

Index